Hotel Law

Transactions, management and franchising

Hotel Law: Transactions, management and franchising presents a practical guide to the issues that face lawyers and industry leaders working in the hospitality field. It aims to develop the reader's understanding of the acquisition process and the complex relationships in management and franchise deals that dominate the hotel industry.

This text is written primarily as a desktop reference for legal practitioners working in the hotel law field and is also suitable for students studying towards hotel and hospitality careers both at an undergraduate and law school or graduate level. The highly experienced author, contributors and editors offer insights into the industry players and their preferred positions, desired outcomes, and the potential pitfalls that can ensnare even the most well-planned deals.

With broad coverage of the rapidly growing field of hospitality law – including gaming, recreation, and amenities – the book's approach examines the dominant models of hotel ownership, management and franchising, and includes independent hotels and the move towards complex resorts. The book's coverage of key legal topics ranges from real estate, to intellectual property, contracts, and finance. *Hotel Law* will give readers an understanding of the hospitality industry from the perspective of the transactional practitioner, while examining the multi-party relationships and agreements that develop between an owner, operator, licensor and lender.

About the author

Nelson F. Migdal has focused his law practice on the hospitality industry for over 30 years. He has a global practice and routinely assists clients with hotel acquisitions, operations, development, hotel-related amenities such as gaming, hotel finance, and hotel management, franchise and license agreements. Nelson is a shareholder at the law firm Greenberg Traurig LLP and co-chairs the firm's global hospitality practice group. He is also a member of the International Society of Hospitality Consultants, President of the Academy of Hospitality Industry Attorneys and Professor in Residence (Adjunct) at the Washington College of Law of American University in Washington, D.C.

About the contributors

Steven Shapiro is Director of the Hospitality and Tourism Law Program at the Washington College of Law at American University, Washington, D.C., where he is also a Professor (Adjunct) in business and real estate topics. With a law degree and a master's degree in civil engineering, Mr. Shapiro started his career as a practicing attorney and is now a project executive for a major construction company. He has been an adjunct professor at the University of Maryland and the Johns Hopkins University.

Tara Gorman focuses her practice on hotel acquisitions, operations, development and finance, hotel management agreements, license and branding agreements, restaurant management agreements, and general commercial real estate office and retail leasing. Tara is a shareholder at the law firm Greenberg Traurig LLP with a concentration on transactions, writes a column for *Hotel Business Review* and is an Adjunct Professor at the Washington College of Law of American University in Washington, D.C., as part of the Hospitality and Tourism Law Program.

Hotel Law

Transactions, management and franchising

Nelson F. Migdal

with contributions from Steven Shapiro and Tara Gorman

Taylor & Francis Group

LONDON AND NEW YORK

First published in paperback 2024

First published 2015
by Routledge
4 Park Square, Milton Park, Abingdon, Oxon OX14 4RN

and by Routledge
605 Third Avenue, New York, NY 10158

Routledge is an imprint of the Taylor & Francis Group, an informa business

Publisher's Note
The publisher has gone to great lengths to ensure the quality of this reprint but
points out that some imperfections in the original copies may be apparent.

British Library Cataloguing-in-Publication Data
A catalogue record for this book is available from the British Library

Library of Congress Cataloging in Publication Data
Migdal, Nelson F., author.
Hotel law : transactions, management, and franchising / Nelson F. Migdal, with
contributions from Steven Shapiro and Tara Gorman.
pages cm
Includes bibliographical references and index.
1. Hotels–Law and legislation–United States. 2. Hospitality industry–Law and
legislation–United States. 3. Hotels–Finance–Law and legislation–United States. 4.
Hotel management–United States. I. Shapiro, Steven, author. II. Gorman, Tara,
author. III. Title.
KF2042.H6M54 2015
343.7307'864794–dc23
2014040259

ISBN: 978-1-138-77952-5 (hbk)
ISBN: 978-1-03-292827-2 (pbk)
ISBN: 978-1-315-77124-3 (ebk)

DOI: 10.4324/9781315771243

Typeset in Bembo
by Cenveo Publisher Services

For Joyce, Talia and Corey with love

'The business of hotel management and franchising is completely interdependent with the law applicable to those endeavors. Though this may be true in all things, these are complex relationships that do not exist in a linear environment. Nelson Migdal has written a comprehensive, though admirably brief, volume that will benefit both legal practitioners and business personnel in the hotel transactions milieu. This volume should be used in every University Hotel Program's curriculum, and it should be in the library of every hotel company, management firm, investor, lender, developer and consultant, so it can, as it should, be read and re-read by hotel group counsel, in-house or outside, and, perhaps more critically, any development/acquisitions/franchise/hotel lending executive working on one of the multiple sides of a hotel transaction, and all of their non-legal advisors.'

Michael C. Shindler, *President,*
Four Corners Advisors, Inc.

'Have you ever found yourself involved in a complicated hospitality industry matter, worrying about whether you were missing something and wishing you had an expert to consult with? *Hotel Law* gives you a thorough overview and makes you feel like Nelson Migdal and his colleagues are right there with you, helping you work through all the issues. I practiced with Nelson for many years. State Bar associations may not allow an attorney to say that they are expert in an area, but I can tell you that the hospitality industry recognizes Nelson as an expert who understands all sides in the industry. He has taken that expertise and made it available to the reader in *Hotel Law*.'

James M. Norman, *Retired Co-Chair,*
Holland & Knight Global Hospitality Resort & Timeshare Group

'Rarely does a legal text come along that offers the technical legal guidance and practical advice on meaningful and topical matters that arise daily in a hospitality legal practice. Whether transactional or day to day operational issues, *Hotel Law* offers the reader context, instruction and validation relating to the various aspects of the lodging and hospitality business. Both in-house and private practice lawyers will find Hotel Law to be a helpful reference tool.'

Terence P. Badour, *Executive Vice President Law & Administration,*
FRHI Hotels & Resorts

Contents

    Representative forms at the companion website.
    Please visit www.routledge.com/cw/migdal to view the forms

    Form of hotel management agreement (provided by
    Fairmont Hotels and Resorts) at the companion website.
    Please visit www.routledge.com/cw/migdal to view the form

    Sample incentive management fee provisions at the companion website.
    Please visit www.routledge.com/cw/migdal to view the examples

    Master date collection checklist at the companion website.
    Please visit www.routledge.com/cw/migdal to view the checklist

    Form of closing checklist at the companion website.
    Please visit www.routledge.com/cw/migdal to view the checklist

Form of franchise agreement (provided by
Marriott International) at the companion website.
Please visit www.routledge.com/cw/migdal to view the form

Form of franchise agreement (provided by
Hyatt Place Franchising, LLC) at the companion website.
Please visit www.routledge.com/cw/migdal to view the form

Form of comfort letter (provided by Marriott International)
at the companion website. Please visit www.routledge.com/cw/migdal
to view the form

Chart of common lender and manager SNDA positions at
the companion website. Please visit www.routledge.com/cw/migdal
to view the chart

Welcome and check-in

By opening to this page you have "checked in" to our hospitality story. We will explore the essential components of hotel acquisition and sale, franchising, and the complex relationship between a hotel owner and its manager that falls under the general heading of hotel management, which includes other agreements that are customarily associated with the management process and various hotel operational structures. Be prepared for some surprises. Some of the concepts, principles, and operating conditions that govern the hospitality industry vary greatly from those of other sectors of the real estate industry. Many knowledgeable and experienced business people and their lawyers and advisors have been shocked and dumbfounded by the significant differences between standard operating procedures for something like an office building and a hotel. For example, experienced owners of retail properties like to analyze a hotel management agreement in the same way that they analyze a lease. Although there are some leases still operative in the United States in the hospitality industry, the analysis of a hotel management agreement is vastly different than that of a lease. Special issues arise when hotels are part of mixed-use projects or when one brand or the operator is replaced by another.

We hope this book will help guide your study and exploration of hospitality industry law practice, or, if you are already a more experienced student or practitioner, that it will deepen your knowledge and experience.

Our plan is to provide academic and practical insights from the perspective of the legal practitioner, with an eye always on both the actual and practical dimension of the law as applied to the industry. We will explore the essential cases that have shaped the industry and the law over the years, and the approaches that may be indicative of how we proceed from here.

We hope you enjoy your stay!

Nelson
Steven
Tara

Acknowledgements

It would not have been possible to write this book without the encouragement, help, and support of family, professional colleagues, and friends from near and far. Many of us who work in the hospitality industry have discovered that solid skill sets and knowledge can best find their voice through solid relationships and professional cooperation.

Many large international hotel management companies are fortunate to have at the helm of their legal departments or within their legal departments some excellent students of hotel law and fine individuals. Edward Ryan, Esq., the General Counsel of Marriott International, Inc. was very supportive of this effort, and with Jane W. LaFranchi, the Assistant General Counsel of Marriott International, Inc., enhanced the materials we could provide to readers. Terrence Badour, Esq., the General Counsel of Fairmont Hotels and Resorts, has been a friend and professional sounding board of ideas over many years. His gracious and knowledgeable support to this effort is deeply appreciated. Margaret Egan, Esq., the Associate General Counsel of Hyatt Hotels Corporation, and John Dent, Esq., the General Counsel of Extended Stay America, were gracious and supportive in permitting the use of some of their brand materials and documents in this book and have participated over many years in expanding and deepening the knowledge base of hospitality lawyers in our industry. Kathleen Barlow and Ariel Silva, both Senior Vice Presidents at Marsh, Inc., provided needed insights into the world of insurance and Laura Mutterperl, Esq., Senior Director and Association General Counsel of Starwood Hotels & Resorts Worldwide, Inc., provided help in understanding brand standards.

My partners at Greenberg Traurig, LLP provided their deep and extensive talents when help was needed. David Oppenheim, Esq., a shareholder in New Jersey, added his franchise knowledge to those elements of this book and made that chapter deeper and stronger. Tara K. Gorman, Esq., a shareholder in Washington, D.C., added the introductory materials, glossary, charts, and other materials to this work. Special thanks to Jean McGruder-Jackson for assisting with presentation of the early manuscripts.

David M. Heller of the American University Washington College of Law provided the initial drafting for the chapter on the essential case law on Management Agreement termination and did an outstanding job of synthesizing the case history into a tight chapter for this book. I hope David continues his interest in hotel law and finds a place in the industry.

I have been fortunate to have two great friends and colleagues in the hospitality industry and study of hotel law. James Norman, Esq., Holland & Knight LLP, retired, and Michael C. Shindler, Esq., the founder and CEO of Four Corners Advisors, have been fellow travelers and management agreement "geeks" with me for decades. They are true students of the law, even though each of them has been at it a long time and is an "expert" in his own right. Jim and Michael provided detailed peer review and insight of the entire book and devoted more time and energy to making it a better work than I would have ever requested. I will be forever grateful for having the benefit of their intelligence and friendship.

Professor Steven G. Shapiro, Esq., Director of the Hotel Law and Tourism Program at the American University Washington College of Law, is primarily responsible for my undertaking of this project and a variety of related academic projects involving hotel law. He has single-handedly built a hotel law program from the ground up and demonstrated a level of commitment and perseverance that is unmatched anywhere.

Undertaking a project such as this requires many things; but most of all, it requires time. In the final analysis, every moment devoted to this book was taken from either rest or family, and, it has to be said, it was mostly family. I hope that Joyce, Talia and Corey are pleased with how this turned out and can take some pride in its completion and publication. They helped me more than they know and I appreciate everything they bring to my life.

1 Evolution of the Hotel Management Agreement

The Hotel Management Agreement is a relatively "new" document in the context of the hospitality industry as a whole. Hotel Management Agreements did not become widely used in North America until the 1970s. In the early days of the hospitality industry, prior to the Second World War, there was no need for close scrutiny of a Hotel Management Agreement because the Owner of the Hotel was also the Hotel Manager, and was either responsible for or personally performed the tasks of the receptionist, the bell staff, the housekeeper, the chief cook and the bottle washer. That was then, this is now.

In today's commercial environment, the ownership, operation, and Branding of the Hotel, as well as its debt financing and capitalization, are in many hands, and each has its own separate and distinct set of interests. Those interests may not be in proper alignment all the time. To set the stage for what is to follow in this book, this chapter reviews the evolution of the hospitality industry, and how that evolution relates to the evolution of the Hotel Management Agreement and the Owner–Manager relationship. This book will examine various aspects of the Hotel Management Agreement and the Owner–Manager relationship. It will explore the specific issues and provisions addressed in the Hotel Management Agreement, as well as the development of specific issues which impact the negotiation, drafting, implementation, and enforcement of the Hotel Management Agreement and the ancillary documents. While the hospitality industry encompasses many different types of hospitality venues and amenities, such as restaurants, recreational resorts, cruise ships, and amusement parks, this book will focus primarily on guest lodging facilities, the legal importance of the Hotel Management Agreement, the Franchise Agreement and the Owner–Manager and Franchisor–Franchisee relationship.

Travel has been a part of life for centuries. As soon as people had the capacity to journey from their homes, villages, and towns, there was the need for a place to stay once the destination was reached. In their infancy, Hotels were simply the home of an innkeeper who offered rooms to the public for overnight lodging. The hospitality industry can thank the Roman Empire for expanding the hospitality business by encouraging the inns and hotels to cater to the pleasure traveler. Evidence of the first inn in America was recorded in 1607, but it was during the Industrial Revolution that the construction of Hotels in Europe, England, and America truly expanded and flourished. From there the hospitality industry took on a life of its own.

The bifurcation of Hotel ownership and operation initially took the legal form and structure of a lease. Hotel leases were much like any other lease of property with a fixed rental rate paid by the tenant to the landlord, and the rent might also have been subject to an annual rental rate increase. This gave the landlord the security of a fixed rental income for the term of the lease, without any benefit or burden dependent upon the profitability of the Hotel. The tenant would take all of the financial risks in connection with the operation of the Hotel. Many Hotel leases structured in this manner eventually evolved to a flat rental rate and a variable or percentage rent based upon a percentage of Gross Operating Revenues or Gross Operating Profits, to allow the landlord the opportunity to share in the profits of the Hotel. The Hotel lease more closely resembled a typical real property lease, and looked very little like the Hotel Management Agreements of today.

Prior to the Second World War, the prevailing structure was that the ownership and operation of the Hotel was held by one individual or one entity, and in many cases, one family. The owner then hired a General Manager to manage the operations of the Hotel on a day-to-day basis. This was a rather simple business arrangement and so too was the operative document, if there were any written agreement at all.

The early Hotel leaders and visionaries of the hospitality industry, such as Kemmons Wilson (Holiday Inn), Conrad Hilton (Hilton Hotels & Resorts), Cecil B. Day (Days Inn) and Bill Marriott (Marriott International) transformed the hospitality industry. In the beginning, these hoteliers owned the real estate, owned the Hotel and Brand, and managed the operation of the Hotel. This resulted in a unity of ownership and operation. There was no need for a heavily negotiated Hotel Management Agreement because the interests of ownership and operation were aligned and unified.

Conrad Hilton opened his first Hotel in Cisco, Texas, in the oil boom times of the early 1930s. The Great Depression was devastating for Hotel Owners, and the hospitality industry might have become extinct, but for the Second World War. The Second World War created a great need for the public to travel, and with travel came the need for transient lodging facilities. Construction materials were scarce during the war, so few new Hotels were built; however, many Hotels that had previously fallen on hard times and closed their doors were reopened to meet the growing need for overnight housing. In fact, the need for transient lodging was so great that many office buildings and warehouses that were closed during the Great Depression were reopened and reconfigured as operating Hotels.

The 1950s ushered in a new era of prosperity and mobility in the United States. With the expansion and growth of the suburbs came the desire of almost every family to own an automobile. The travel industry took a new turn, driven by accessibility to the automobile, the growth of air travel, and the financial capacity to travel. The traveler was now driving to tourist courts, roadside inns, bed and breakfasts and Hotels not only in the cities, but also in the countryside. The "tourist courts" became "motor hotels," which we now call "motels."

With the mobility of the leisure traveler came the need to provide predictability in transient lodging facilities. To meet the increased demand Conrad Hilton expanded his operations and real estate holdings. But it was Kemmons Wilson who capitalized on the need to provide predictability. Kemmons Wilson franchised the first Holiday Inn in 1957. Wilson's original goal was that his Hotels should be standardized, clean, predictable, family friendly, and readily accessible to road travelers. By 1958 there were 50 Holiday Inns across the country, all of them fulfilling his criteria exactly. The idea of predictability revolutionized the hospitality industry and laid the foundation for the development of a brand. By the 1960s Holiday Inns and similar Brands replaced most of the pre-war tourist courts. During the 1960s, hoteliers quickly discovered that the key to growing their respective Brands was a combination of public ownership and Franchising.

The legal structure of a Franchise Agreement brings advantages to Owner of the Hotel in the form of access to the Brand's marketing and sales force, reservation services, and a strong Brand name when soliciting financing. Yet franchising allows the Hotel Owner to maintain independence in the operation of the property. Owners might choose to operate the property themselves, while others might choose to hire a Third-Party Manager to operate their property.

The 1960s also brought new owners into the industry, such as life insurance companies as Owners of transient lodging facilities. The core competency of life insurance companies is clearly not operating Hotels. For that reason, the life insurance companies partnered with the hoteliers to operate the hotels. This enabled the owner/managers to sell the asset to the life insurance company and lease the asset back thereby freeing up cash for the owner/managers and separating the ownership function of the Hotel from the operation function of the Hotel.

The separation of the ownership function of the Hotel from the operation function of the Hotel in the late 1960s to early 1970s was truly a transformative moment for the hospitality industry. The life insurance companies would own the asset and the hoteliers would operate the asset. This, in turn, led to changes in the Hotel Management Agreement. The relationship between Owner and Manager was no longer simple, so a simple operative document would no longer suffice. The Hotel Manager was no longer a family member, tenant, or "at-will" employee, but rather another legal entity, often with a Brand affiliation. The change in the ownership structure resulted in a change in the way hotels operated.

The days of the grand hotelier and the unity of ownership and operation are all but gone. Hotel Management Agreements became widely used in North America in the 1970s. The bifurcation of ownership and operation as set forth in the Hotel Management Agreement creates two power centers that require initial balancing and continual rebalancing during the term of the relationship of the parties. In the early days of Hotel Management Agreements, the Brand, with its operational expertise and Brand affiliation, held the majority of the negotiating power and control over the process. Often, the individual Owner had little to no hospitality industry knowledge or experience, resulting in Hotel Management Agreements that were favorable to the Manager. During strong economic times or times when significant tax advantages and strong profitability find both parties earning a profit and benefitting from the relationship, not much attention is paid to the balance of power and to which party may be taking the greater financial risk. When President Gerald Ford signed the Tax Reform Act of 1976, which included the initial legislation providing enhanced benefits to the Real Estate Investment Trusts (REITs) structure, the foundation was established for a strong REIT presence in hotel ownership and the potential for significant growth of the industry. On the other hand, during weaker points in economic cycles, any imbalance of power becomes more evident. Over time, the character and structure of Hotel Owners also changed; they became more sophisticated and began paying greater attention to how Hotel Management Agreements impacted the value of their Hotel. Moreover, they came to appreciate the value in maintaining a certain degree of control over their Hotel to protect their investment.

Today, most Hotels are owned by one party, Branded or Franchised by another, and, in some cases, managed by a third entity, all of whom must answer to the Lender in one form or another. Today Hotel Owners might be structured as corporations, limited partnerships, limited liability companies, or publicly held entities in the form of REITs.

Family-owned Hotel chains and owner/manager entities continue to exist, but that is no longer the ownership structure for most Hotels in the United States.

Management of the asset is often completely separate from ownership of the asset, and Hotel Owners, whether high net worth individuals, lenders, institutional investors, pension funds, private equity funds, REITs, or any other legal structure, are focused on the quality of their investment and its revenue-generating capacity. For many in the hospitality industry today, the Hotel asset is just another asset class in their portfolio, part of a diversification strategy that may include retail, office, residential, industrial, and commercial properties. Today's Hotel Owners are certainly more sophisticated and knowledgeable about the economics of the hospitality industry than the Hotel Owners of the 1970s, although it is sometimes unclear whether they are more knowledgeable and savvy about the "hospitality" part of the business and guest-facing amenities. Many Owners also engage advisers, asset managers, and lawyers to protect their interests, transforming the once simple Hotel lease into the sophisticated and complex Hotel Management Agreement.

This book is not intended to be a survey of every element of what might be considered part of Hotel law or event hospitality law. By concentrating on the acquisition and sale process, franchising, and management, there will be opportunities for in-depth analysis of some of the common positions and arguments in transactions and making new law today. The desire and intent is to strike a balance and offer insights into the thinking of sellers, buyers, franchisors, franchisees, owners, managers, and lenders, and how the balance of power is reflected in the documents and structures in each subject area. In some situations, such as describing hotel management by Brands or franchising, where the initial documents for the Owner–Manager or Franchisor–Franchisee relationship are produced by the Manager or Franchisor, the only way to examine the issues and dynamics is in reaction to what is presented to Owner or Franchisee. The use of the "Owner's voice" in illuminating the hotel management or franchise documents arises out of necessity, but, nevertheless, with the intent of fairly describing the competing interests of the parties.

2 Hotel Acquisition and Sale

Overview

A sophisticated real estate investor may know all the ins and outs of purchasing office buildings, retail venues, residential complexes, and mixed-use properties, but may fall short when it comes to identifying and avoiding potential landmines in Hotel acquisitions and sales. While purchase agreements used in Hotel acquisitions may appear familiar at first glance, there are several key considerations that both sellers and purchasers should appreciate before undertaking a Hotel acquisition or sale, including a site for future hotel development or mixed-use including a hotel. Since every acquisition by a purchaser requires a sale by a seller, this chapter will provide some insight into things to look for during the entire Hotel sale and acquisition process.

The hotel as an operating entity

The acquisition or disposition of a Hotel involves an operating business that takes no holidays and does not get a day off. Every moment of every day, every day of the year, the Hotel is open and operating. Employees are going about the business of Hotel operations and the delivery of services and amenities to guests, and Hotel guests are engaging in a wide variety of activities depending upon the service level, location, and size of the Hotel. The sale and purchase of that kind of an asset must account for all of those moving parts, avoid disruption of ongoing activities, and do minimal or no harm to the guest experience. In many ways, a Hotel is like a transitional city, where the residents change daily, but the government stays in place.

Flag or no flag

There are many very familiar "Brand" names of Hotels and national and international Hotel operating companies. Their names and familiar logos and signs are meaningful to potential Hotel guests and carry value that Hotel industry investors and loan underwriters recognize in Hotel valuations. Whether a Hotel carries a "flag," "Franchise," or "Brand" is an important factor for prospective sellers and purchasers to consider in the universe of transactions representing components of what we consider the Hotel sale and purchase. There are certainly un-Branded or independent Hotels flourishing today. Some Hotels are historic or have a special and unique story that sets them apart from the competition. These independent Hotels have enjoyed an additional lift in popularity as so-called "lifestyle" and "boutique" Hotels continue to gain popularity with certain age groups, in certain cities and at certain price points. Aside from that group of independent Hotels, the presence of a flag or Brand impacts many aspects of the transaction when a Hotel is about to change hands. The financial health of the Hotel today and in the future as a result of the professional management of day-to-day operations of the Hotel, the inventory of personal property that is sold, along with the physical structure of the Hotel, collective bargaining agreements with unions, unfunded pension liability, and intellectual property rights are all impacted.

The Brand of the Hotel is critical to the revenue generated from the operation of the Hotel and the overall value of the Hotel. For example, if a Hotel changes its Franchise affiliation and goes from a strong Franchise to a weak Franchise or to no Franchise, the Hotel may lose a portion of the revenue stream that might have come through a strong central reservations system, the Brand's rewards program, and an array of amenities provided by the Brand through what are commonly referred to as "centralized services." A Hotel that is sold "encumbered" (in the language of Hotel Owners) or "enhanced" (in the language of Hotel Operators) by a strong flag may be valued much higher than an independent Hotel or a Hotel that is sold with a weak flag, unless the Hotel has other features, which may simply be its location, to give it strategic value. Purchasers must consider current management or Branding of the Hotel along with the physical structure of the Hotel, comprised of the land and the building, when formulating purchase and sale decisions and documents. Part of what is being purchased may be Manager or the right to use the name of the Hotel, and all the benefits and obligations attendant to third-party hotel management or franchising. Sellers must ponder the same consideration, along with one other: the potential that the Hotel Management Agreement or the Franchise Agreement includes a right of first offer, right of first refusal, or some other purchase right or option for the benefit of Manager or Franchisor. These types of rights on behalf of the Hotel Manager or Franchisor are indicative of a Hotel that may serve a geographic or other strategic need within the System of the Hotel Manager or Franchisor. The challenge for Seller is the potential that this type of right, depending upon how it is expressed, can have a chilling effect on the sale process and delay the sale of the Hotel. Hotel Seller should not assume that the sale of the Hotel can occur without the involvement of Manager or other advisor as a matter of right. If the Seller does not have an Asset Manager that may be intimately familiar with the Management Agreement or Franchise Agreement, the Seller needs to check for purchase options as part of the initial selling strategy.

A Branded Manager's desire for an opportunity to purchase the Hotel arises from the long-term nature of Hotel Management Agreements and the belief that the Branded Hotel may be a strategic component of Manager's larger collection or system of Hotels. Under many Management Agreements, the sale of the Hotel does not result in the termination of the Management Agreement, and Manager stays in place, serving a new Hotel Owner. There are other situations in which a sale of the Hotel includes the right to terminate the Management Agreement in exchange for the payment of a significant termination fee, or, occasionally, no fee at all. If the Hotel is strategically situated, the sale of the Hotel with the termination of the Management Agreement may be a significant blow to the footprint of the Brand, leading the management company to include a right to purchase the Hotel in the Management Agreement. Similarly, many franchised Hotels may be strategically located and important to Franchisor. Consider urban markets with high barriers to entry such as New York, Chicago, Los Angeles, Boston, San Francisco, Washington, D.C., and Miami. The loss of Brand presence in locations with those characteristics will be a strong motivation for the Franchise Agreement to include an opportunity to acquire the Hotel. Another variation on the dilemma of pre-serving Brand presence is the Management Agreement with the potential of conversion to a Franchise. This is sometimes described as a "springing franchise," but the general scope of the process is a Management Agreement that the Hotel Owner can terminate, so long as the termination occurs simultaneously with the Hotel Owner entering into the Brand's standard Franchise Agreement, usually, although subject to negotiation, for the remainder of the term of the Management Agreement or the customary term of a new Franchise Agreement, whichever is longer. Both the Hotel Manager and the Hotel Owner can benefit from this opportunity. The Hotel Manager knows that its Brand will retain a presence at the Hotel for a long period of time in one form or another, while the Hotel Owner has an option to convert the Hotel from Brand Management to Third-Party management, so long as the Brand stays in place.

The concept of a Branded or Brand Manager is intended to describe the situation when the Brand is also Manager of the Hotel. The Hotel Owner engages the Hotel Manager to manage the Hotel in accordance with the Brand's System and Brand Stand-ards. The Branded Manager does not merely license its Brand, including intellectual property, logos, goodwill, and systems to the Hotel Owner. Instead, the Brand Manager manages the Hotel in accordance with its own systems that the Hotel Manager delivers on behalf of Owner. This should be understood and distinguished from a license arrangement or a franchise agreement under which a Brand permits a Hotel Owner to use the Brand's Systems. In that situation, the Brand is permitting another party, the Hotel Owner, to use the Brand's intellectual property, logos, goodwill, and systems, but day-to-day management of the Hotel is in the hands of Owner or a Manager-approved third-party Manager that is not the brand itself.

This will play out in the negotiation of a Purchase and Sale Agreement ("Purchase") through a clear understanding of the terms and conditions of the Franchise Agreement or Management Agreement, the ability to terminate that agreement, and the elements of the asset and the operations that are encompassed within such an agreement. Specifically, it is a common oversight to see an Agreement that defines the "Property" being sold and conveyed by the seller to include rights and duties relating to employees, permits and licenses, and marks and Brands used in commerce to identify the Hotel. With a "Flag" in place, all of those rights are generally owned and controlled by the Hotel Manager and cannot be sold or conveyed by the Seller. This also means that the Hotel Manager will have the ability to protect its property and its agreements with the Seller by having the right to approve the new owner of the Hotel (the Buyer) and particularly insure that the Buyer will not be a competitor of the Hotel Manager. The Hotel employees can create a similar concern because they are often employed by Manager and not the Hotel Seller. The operating permits, particularly the liquor license, are often in the name of Manager. The proprietary marks are almost always the sole and exclusive property of Manager or its parent company.

This is an important "lesson to be learned." Assume nothing. The initial due dili-gence for both seller and purchaser should involve an assessment of who owns what and if and how it can be transferred or conveyed. This should come before the development and legal teams spend too much time developing the checklist of what the seller is expected to sell under the Agreement. The seller may not actually own everything it thinks it is trying to sell, and the purchaser may be negotiating with the Hotel Manager or Franchisor as well as with the seller.

(To more vividly describe the importance of due diligence and achieving a full comprehension of all aspects of the sale and purchase transaction, we have added examples of master data collection and due diligence checklists at the companion website. Please visit www.routledge.com/cw/migdal to view the examples.)

Franchise Agreements and License Agreements

The Hotel purchase and sale transaction requires a significant amount of study and planning. This is generally referred to as the purchaser's "due diligence". It is the research, study and analysis conducted to make sure that the purchaser desires to proceed with the transaction, and that the purchaser can use the property in the manner it expects. During the due diligence period, purchasers must examine all contracts, supply and service agreements and ongoing purchase orders and agreements that may be transferred from seller to purchaser in connection with the right to use and affiliate with the Brand. These rights may come in the form of a Franchise Agreement, License Agreement, Branding Agreement, or through a Hotel Management Agreement.

Traditionally, within the United States, Branded Hotels have operated through Management Agreements or Franchise agreements, whereby Owner of the Hotel acquires the right to use the name of the Hotel and other benefits associated with the Brand in exchange for the payment of monthly fees, or the right to have the Hotel managed by Franchisor in exchange for base fees and incentive fees. There may still be lease relationships in the marketplace, particularly operating leases to accommodate REIT or other ownership structures. But essentially, a transaction to sell and buy a Hotel that is not an independent hotel will involve a hotel that is either managed by the seller, a company affiliated with the seller or a third-party management company under a Franchise Agreement or be subject to a Management Agreement. The more common arrangement is for the Hotel Owner to obtain the Franchise and be Franchisee, and then Owner hires a Manager or Operator to Manage the Hotel in accordance with the Franchise Agreement and the Management Agreement. The benefit of this structure is that Manager hired by Owner is obligated to maintain the operating standards set forth in the Franchise Agreement. Franchisor will retain the right to approve Manager and, in most instances, the Management Agreement with Owner. The significant point to remember is that when an Owner hires the Brand as its Manager, rather that becoming a Franchisee under a Franchise Agreement, the primary responsibility for day-to-day operations resides with Manager. Manager must manage to the operating standards of the Brand, but at the expense of Owner. Manager will not be in default for failure to properly operate the asset if that failure is the result of or arises from Owner's failure to fund operations as required by the approved Annual Budget. The decision to proceed under a Management Agreement or under a Franchise Agreement is a crucial one. Not every Brand is available as a Franchise, which creates an inherent limitation on an Owner's choice. At the upper end of the segmentation of the hospitality business where the Hotel is considered upscale or luxury and excellence in the delivery of guest amenities and services is of paramount importance, many Brands require management by the Brand and will not permit a Franchise. When either management or franchise is available to a Hotel Owner, it will undertake a cost-benefit analysis comparing management by the Brand against a Franchise Agreement with a third-party management structure overlaid on top of the Franchise Agreement. A true comparison based purely on fees is challenging; although many publications produce schedules of fees that can be assessed by most

Franchisors (usually stated in terms of maximums set forth in Franchisor's disclosure documents), the fees assessed under a Hotel Management Agreement are not publicly available in a similar fashion. Nevertheless, some Hotel Owners prefer to engage a Third-Party Manager to manage the Hotel pursuant to a Franchise Agreement and a Management Agreement rather than engage the Brand as the Hotel Manager.

As a way to differentiate a Hotel from the competition, high-end hospitality companies and even celebrities have been licensing the right to use their Brand name to Hotel Owners. This right comes in the form of a License Agreement or a Branding Agreement with very specific Brand standards that Owner of the Hotel must comply with in order to maintain the right to use the Brand. In this case, Owner purchases the right to use the Brand name of the licensor but is not required to have the Hotel managed by the licensor. The pitfall here is that in the event the Hotel is not managed in compliance with the Brand standards, the licensor can terminate the license agreement and pull the Brand from the Hotel. More often than not the strength of the Brand is a critical component when obtaining Hotel financing, and if the license agreement is terminated, Owner may be immediately in default with its Mortgagee, or a default is most probably imminent.

With food and beverage services becoming such an important differentiator of hotels at almost every full service segment of the business as well as some limited service offerings with food and beverage available at a site adjacent to the Hotel, the Hotel Owner might find itself with a separate lease, management agreement, branding or license agreement or some variation on these themes to weave into the entire fabric of hotel operations and guest amenities. Food and beverage operations can bring additional revenue to the Hotel and be a big value-added for ownership, but it can just as easily be a consistent loss leader. This potential amenity will require its own due diligence and analysis to insure that the potential rewards support the additional complications.

Special lender rights and mortgagee comfort

Although the number of Hotel transactions completed on an "all cash" basis without Owner obtaining a loan has increased, the more common scenario is for the purchaser to contribute some equity to the project to better manage leverage, but then complete the purchase with debt. In the transaction for the transfer of a Hotel from seller to purchaser and purchaser's debt financing of the acquisition, Mortgagees will require a comfort letter from Franchisor or, in the management environment, a Subordination, Non-disturbance, and Attornment Agreement ("SNDA") as part of the purchaser's financing. The purpose of the comfort letter is to provide the Mortgagee "comfort" that in the event the new Owner (its borrower) is in default under its loan documents, Mortgagee can step into the shoes of its borrower (Owner), cure the default and assume the obligations under the Franchise Agreement in order to keep the Brand with the Hotel. Because the Brand carries value, the Mortgagee needs the comfort that Franchisor will send the Mortgagee default notices upon a default under the Franchise Agreement and recognize Mortgagee as Franchisee under the Franchise Agreement, if necessary. Many Franchisors keep all of their options open, and reserve to themselves the right to terminate the Franchise Agreement, or, at least, charge the new Owner an application fee or other form of transfer fee in connection with the transfer of the Franchise Agreement to a new Owner. Recently, some Franchisors have been approaching the comfort letter as if it were an SNDA in the sense that in the event of a borrower default under the loan documents, Franchisor will deny Mortgagee the right to terminate the Franchise

Agreement and instead Mortgagee must choose between the assumption of the existing Franchise Agreement or the execution of a new Franchise Agreement. This can prove challenging for many Mortgagees who want the unfettered right to terminate the Franchise Agreement if the borrower defaults and the Mortgagee takes Ownership of the Hotel through foreclosure or a deed in lieu of foreclosure.

The SNDA is vastly more complicated because Manager must preserve the right to continue to operate the asset, receive at least the base fee, and essentially not be terminated by Mortgagee or any subsequent Owner through foreclosure solely based upon Owner's default under its loan documents. It is essential to Manager in light of the long-term nature of the underlying Management Agreement that Manager remain undisturbed in its possession and management of the Hotel.

From the perspective of the Purchase Agreement, the purchaser must anticipate the need for a comfort letter, SNDA or both to close its loan, and the likelihood that the Hotel Management Agreement may already stipulate a form of SNDA or the requirements of the SNDA, just as each Franchisor will have its own form of comfort letter. The sooner this aspect of the transaction is examined and resolved, the better it will be for both seller and purchaser.

Property Improvement Plan

Purchasers should not be complacent about the transfer of a Franchise or assignment of a Hotel Management Agreement upon sale simply because the agreement is in place and in effect. Franchisors and Managers often use the transfer of the Hotel to a new Owner as an opportunity to upgrade the Hotel. Franchisor will provide the new Owner with a Property Improvement Plan ("PIP"). The PIP is a plan for renovation of the Hotel which will generally set forth a list of Hotel improvements Franchisor will require Owner to make if the Hotel is to maintain the Brand. Initially, this "facelift" for the Hotel may seem like a good idea to give the Hotel a fresh start with the new Ownership; however, the implementation of the PIP may come at a significant cost, which is frequently not part of the purchaser's initial calculation of acquiring the Hotel. In addition, there might be a PIP in progress at the time the Hotel is changing hands. The PIP itself is usually the schedule of renovations and repairs. It does not set forth the cost. It becomes incumbent upon the careful purchaser to dig deeper into all aspects of the PIP and all potential impacts on the Hotel. This may involve an examination of how seasonality impacts the Hotel in light of the timing of the PIP; impact on group and individual bookings; the time leading up to commencement of the PIP; the time to complete the PIP; the impact on the structure of the building and any major mechanical or operating systems; and Americans with Disabilities Act compliance and general local law compliance, so that the purchaser can factor the PIP and all PIP-related costs into its debt and equity structure. The wise purchaser would examine the elements of the PIP to determine if there are any elements that are not related to preserving Brand standards, insuring compliance with applicable laws or protecting guests and their property, and therefore might be phased or staged to better manage purchaser's early cash flow. Ideally, the PIP would be negotiated before the Purchase Agreement is finalized. If that is not achievable, then finalizing the PIP might be a condition precedent to Purchaser's obligation to close. Of course, each negotiation is unique, and there is always room for variations to anything. For example, we have seen situations in which the Purchaser would have the right to terminate the Purchase Agreement if the cost of the PIP exceeded a pre-negotiated maximum dollar amount.

Hotel Management Agreements

The day-to-day operations of the Hotel are far more cumbersome than the day-to-day operations of other real estate investments. For example, a typical office lease runs for a significant term of years that might be five, ten or more years, may contain renewal terms, and contains obligations for the tenant to maintain its premises. This guarantees Owner an income stream and a well-maintained property for many years. Hotel rooms are re-priced and sold every day. Housekeeping and maintenance of the physical plant are constant and require diligent management of both the maintenance process and the people performing the maintenance. More often than not, Owner of the Hotel does not want to take on the day-to-day operations of the Hotel. This has never been more true than it is today with the universe of Owners including life insurance companies, public and private Real Estate Investment Trusts ("REITs"), pension funds, and other private and public entities with investors or shareholders demanding economic results but being poorly equipped for the tasks of management. Even Owners who can be cynical over long-term Management Agreements and relentless in their scrutiny of their Managers rarely decide to self-manage their Hotels and engage a major Brand under a Franchise or under Management for that purpose. The message seems to be that third-party management has a place in the hospitality industry today, even with a critical and demanding Owner.

Hotel Management Agreements cover every aspect of the management of the Hotel from employees, reservations, general maintenance, and capital improvements to books and records and myriad other accompanying issues. During the due diligence period, the purchaser must carefully review the existing Hotel Management Agreement and everything the purchaser will inherit on the day it acquires the Hotel.

Employees

One of the most significant issues for Hotel purchasers is to determine the identity of the employer. Is it Owner (the seller) or Manager? There are pros and cons to each scenario, and the terms and conditions of the Agreement with respect to employees and employment matters must be complete under any circumstances. Typically, Owner is responsible for the costs of employment as an operating expense of the Hotel regardless of which party is the employer. Similarly, the issues of imputation of liability for the conduct of certain employees are often the same regardless of who the employer is. What tips the scale will vary based on many factors, and the applicable law of the jurisdiction in which the Hotel is located. It is not unusual for Manager to be the party best equipped to deal with the legal aspects of employment issues, including collective bargaining agreements. Although it can be adequately addressed in the Agreement, particularly when the current Manager is the employer, the purchaser must be mindful of how the transfer of Ownership of the Hotel might trigger the applicability of the Worker Adjustment and Retraining Notification Act (WARN Act). The WARN Act protects workers against the adverse effects of plant closures by requiring employers to provide notice 60 days prior to Hotel closings or mass layoffs. If the Hotel will continue to operate, Manager is the employer and Owner assumes the Hotel Management Agreement on the Hotel acquisition, and the employees stay in place. However, upon the transfer of Ownership of the Hotel, if the Hotel employees are, generally, full-time employees of the seller, on the date of closing the employees of seller become the employees of purchaser or an entity designated by the purchaser. This may involve Seller terminating the employment of the employees to allow purchaser to then hire them.

There are many issues related to employees when transferring Ownership of the Hotel. For example, the Agreement should address the allocation of all accrued but unpaid employee salaries, wages, bonuses, accrued vacation, profit sharing, and other benefits. This can be a hidden cost if not addressed during the Hotel acquisition process.

Food and Beverage Operations

Food and beverage operations are an important component of a Hotel, whether the Hotel simply has a minibar in each room, a full service catering and banquet operation or in-room dining. Licensing for the sale of alcoholic beverages is a complex matter that is regulated on a state-by-state basis. Although the issues surrounding liquor licensing defy simplification, there are certain basic conceptual matters that are worthy of consideration. The first line of inquiry is which party holds the liquor license or other licenses necessary for the food and beverage operations. If Manager holds the licenses and the purchaser assumes the Hotel Management Agreement or food and beverage Management Agreement, this transition should go smoothly. If the seller holds the licenses, the Purchase Agreement should address the transfer of the licenses. The Purchase Agreement should require the parties to cooperate in the effort to transfer the licenses and that, in the event that the licenses cannot be transferred prior to the date of closing, the purchaser may operate the Hotel using the existing licenses for a period of time. Typically, the Agreement will require a beverage management agreement in connection with the continued use of the licenses by purchaser. This is one area of the Agreement that depends almost entirely upon local law. It is essential that all of the local rules be understood. The opening of a full service Hotel without a liquor license does occur, but it is not a very pretty situation if alcoholic beverages cannot be served.

Transfer of inventory and reservations

The transfer of inventory upon the sale of a Hotel can be quite an undertaking. For example, if the Hotel is Branded, much of the consumable inventory (toiletries, towels, napkins) contains the Brand logo and even the specific location of the Hotel. If this inventory is not transferred to the new Owner and the Brand is changing, license and management agreements typically require that all the inventory be destroyed. Often in Purchase Agreements in jurisdictions that have bulk sales laws applicable to Hotels, the parties expressly agree not to comply with and agree to waive any statutory bulk sale requirements that are applicable to the sale of a Hotel, and instead contractually agree to address the bulk sale and the transfer of inventory in the Agreement. If statutory bulk sales laws do not apply, or if seller and purchaser agree to waive such laws, then the purchaser's protections will reside exclusively within the Agreement. This is generally addressed through seller's representations and warranties, the effects of which survive closing for some period of time, and the breach of which permits purchaser to seek recourse against seller. The Agreement might stipulate an economic threshold below which purchaser cannot pursue seller, as well as a limitation or "cap" on seller's liability. In addition, there may be provisions obligating seller to preserve its corporate integrity or post a cash escrow or other security during a survival period.

Advance reservations for Hotel rooms, meetings, special events and banquets must be addressed in the Agreement. Reservations and deposits must be transferred to the purchaser upon the acquisition of the Hotel and the seller must agree to cooperate with purchaser and act in good faith in the taking of reservations prior to the transfer of Ownership.

Other considerations and sample contract provisions

Hotel Management Agreement

To address the important element of purchaser being approved by the current Manager and assuming the Management Agreement, a purchase and sale agreement might include a provision along these lines:

> The Property is subject to the Hotel Management Agreement with Manager. The obligations of Seller to close the sale contemplated hereby shall be conditioned upon Buyer complying with Section __ of the Hotel Management Agreement, including but not limited to Buyer providing a subordination, non-disturbance and attornment agreement satisfactory to Manager and Mortgagee in accordance with the Hotel Management Agreement. Buyer represents and warrants that Buyer has reviewed the Hotel Management Agreement and will be able to satisfy such requirements. If Buyer does not satisfy the requirements set forth in Section __ of the Hotel Management Agreement and due to such fact, Manager will not permit an assignment of the Hotel Management Agreement to Buyer, then Buyer will be deemed to be in default hereunder and Seller may terminate this Agreement and retain the Earnest Money as liquidated damages.

Observe that Buyer has an obligation to comply with certain provisions of the Hotel Management Agreement. One reason for this is that a sale of the Hotel that is not in conformance with the requirements of the Hotel Management Agreement will constitute a default by Owner (seller) under the Hotel Management Agreement. Although seller is seeking the sale of the Hotel to a purchaser, the seller must also be careful to involve Manager so as to avoid a default under the Hotel Management Agreement.

In this sample clause, Buyer has agreed to satisfy the requirements of the Hotel Management Agreement, and if Buyer does not do so, and due to such fact, Manager will not permit an assignment of the Hotel Management Agreement to Buyer, then Buyer is in default under Agreement, giving rise to seller's right to terminate the Agreement and retain Buyer's deposit as liquidated damages. This puts a great deal at stake for the Buyer.

Liquor licenses

A typical Agreement clause might look like this:

> Buyer acknowledges that there may be various liquor licenses associated with the operation of the Hotel. As soon as is reasonably practicable after the full execution and delivery of this Agreement, Buyer shall file an application with the required state and local agencies (the "Board") for the issuance as of the Closing Date of such licenses (the "Required Liquor Licenses") as are necessary for the continued

operation of the Hotel. Buyer shall diligently pursue the obtaining of the Required Liquor Licenses at Buyer's sole cost and expense. The issuance of the Required Liquor Licenses [shall] [shall not] not be a condition of Closing. If applicable, Seller shall enter into, or shall cause the holder of the liquor licenses to enter into, a Post-Closing Interim Agreement for use of Seller's liquor licenses.

Liquor Licenses always require special care and attention. The application that will be filed will include detailed personal information, such as fingerprint cards and sufficient personal background information for a criminal background check. Upon the sale of the Hotel, the new Owner will be in a difficult position if it cannot serve alcoholic beverages, with very limited exceptions for "dry" jurisdictions. This issue is completely dependent on local law, but the parties to the Agreement must discuss and determine whether the issuance of a liquor license to Buyer will be a condition precedent to Buyer's obligation to close, and, if permitted in the jurisdiction, the manner in which seller will permit buyer to operate under seller's liquor license. Key elements of this discussion are the cost to buyer, buyer's indemnification of seller, insurance requirements, and how long the use will be permitted.

Hotel-specific allocations and pro rations

Because the sale and purchase of a Hotel involves an operating entity, the purchase and sale agreement will address the allocation and apportionment of certain operational income and expense items. This may involve all of the following.

Hotel reservations and revenues

RESERVATIONS

A typical Agreement clause might look like this:

> On the Closing Date, Seller shall request that Manager provide Buyer with its schedule of confirmed reservations for dates subsequent to the Closing Date, which schedule shall list the party for whose benefit the reservation was made, the amount of deposit thereunder, the amount of any room rental deposits, and the amount of any other deposits made for advance reservations, banquets or future services to be provided after the Closing Date, together with complete copies of all relevant contracts and agreements. Buyer will honor (or cause its Manager to honor), for its account, all pre-Closing Date reservations as so confirmed by Seller for dates subsequent to the Closing Date at the rate or price previously agreed to by Seller (so long as such rates conform to customary rates charged by Seller). Seller shall pay or credit to Buyer the amount of all prepayments or deposits disclosed in such schedule.

Observe the impact of this provision. Seller would commonly have represented to Buyer that Seller has operated the Hotel, and caused its Manager to operate the Hotel in the ordinary course of business up until the closing date. Buyer and its Manager, whether it is the prior Manager remaining in place or a new Manager, must honor existing reservations. To the extent that Seller already received deposits or payment for reservations that will occur after the closing, Buyer receives a credit against the purchase price because it is the Buyer who will be executing or discharging the reservation. These types of provisions generally are tied to other provisions for a "true up" or reconciliation within a pre-negotiated period of time after the Closing Date.

GUEST REVENUES

A typical Agreement clause might look like this:

> Revenues from guest rooms in the Hotel occupied on the night containing the Cut-Off Time, including any sales taxes, room taxes, and other taxes charged to guests in such rooms, all parking charges, sales from minibars, in-room food and beverage, telephone, facsimile and data communications, in-room movie, laundry, and other service charges allocated to such rooms with respect to the night containing the Cut-Off Time shall be divided equally between Seller and Buyer; provided, however, that to the extent the times at which food and beverage sales, telephone, facsimile or data communication, in-room movie, laundry, and other services are ordered by guests can be determined, the same shall be allocated between Seller and Buyer based on when orders for the same were received, with orders originating prior to Cut-Off Time being allocable to Seller, and orders originating from and after the Cut-Off Time being allocable to Buyer. All other revenues from restaurants, lounges, and other service operations conducted at the Property shall be allocated based on whether the same accrued before or from and after the Cut-Off Time as described in the preceding sentence, and Seller shall instruct Manager, and Buyer shall instruct its Manager, to separately record sales occurring before and from and after the Cut-Off Time. The foregoing amounts are referred to collectively as "Guest Revenues."

The concept of a "Cut-Off Time," sometimes also called the "Outlet Closing Time" in Hotel transactions, recognizes that the Hotel will continue to have items of income and expense as Ownership is changing hands and as seller and buyer are attempting to prepare accurate closing adjustments. The typical Cut-Off Time is 11:59 p.m. on the date immediately prior to the closing date. The party that will receive the benefit of the income is intended to be the party that also bears the expense of generating that income.

BANQUET AND MEETING ROOM REVENUES

A typical Agreement clause might look like this:

> Revenues from conferences, receptions, meetings, and other functions occurring in any conference, banquet or meeting rooms in the Hotel, including usage charges and related taxes, food and beverage sales, valet parking charges, equipment rentals, and telecommunications charges, shall be allocated between Seller and Buyer, based on when the function therein commenced, with (i) one-day functions commencing prior to the Cut-Off Time being allocable to Seller; (ii) one-day functions commencing from and after the Cut-Off Time being allocable to Buyer; and (iii) multi day functions that include periods both before and after the Cut-Off Time being prorated between Seller and Buyer according to the period of time before and from

and after the Cut-Off Time. The foregoing amounts are referred to collectively as "Conference Revenues."

Here again note the intention that the party that will receive the benefit of the income is intended to be the party that also bears the expense of generating that income.

Unredeemed gift certificates and vouchers

A typical Agreement clause might look like this:

> Buyer shall receive a credit against the Purchase Price at Closing for the face value of all paid for vouchers, gift certificates, and other promotional materials (together, the "Vouchers") which may be used as full or partial payment for any Hotel service including room rentals, food and beverage service, or any other item either borne directly by Owner of the Hotel or which is reimbursable by Owner of the Hotel (i.e., if a gift certificate can be used to pay for items in the Hotel gift shop). The parties also agree that no credit shall be given for any complementary Vouchers. Seller shall request that Manager deliver to Buyer one (1) business day prior to the Closing Date a list of all such Vouchers.

Gift cards and vouchers can be a problem that should be remedied by disclosure and a credit to Buyer of the value of gift cards and vouchers that Buyer may be obligated to honor once Buyer is the Hotel Owner. The most frequent challenge in this area is the failure of many Hotel Owners to track and monitor gift cards, vouchers, and similar cards that are issued. Owners often simply do not know the value of outstanding cards and vouchers that the new Owner will be asked to honor, making it incumbent upon both parties to discuss and settle the matter in the Agreement. While often more manageable, if the Hotel is subject to an agreement with a Brand, a Buyer will want to investigate continuing obligations on the Brand's loyalty or rewards programs.

Hotel Management Agreement indemnity

A typical Agreement clause might look like this:

> Buyer shall indemnify, defend and hold harmless Seller from and against any claim by Manager (and all obligations, claims, liabilities, damages, losses, cost or expenses, including reasonable attorneys' fees and court costs, resulting therefrom) by reason of a default by Owner [which is the Buyer under the Purchase Agreement but will be Owner after Closing] under the Hotel Management Agreement occurring on or after the Closing Date. Seller shall indemnify, defend and hold harmless Buyer from and against any claim by Manager (and all obligations, claims, liabilities, damages, losses, cost or expenses, including reasonable attorneys' fees and court costs, resulting therefrom) by reason of a default by the Seller under the Hotel Management Agreement occurring prior to the Closing Date. The indemnity obligations set forth in this Section shall survive the Closing.

The Closing Date is the turning of the page for the ownership of the Hotel. Buyer remains the primary obligated party but only on and after the Closing Date. Prior to that time, Seller remains obligated for the proper discharge of its obligations. The challenge here is how to support the surviving obligations of a seller, who will have sold the Hotel and disbursed all net settlement proceeds to its Owners and stakeholders. This aspect of the Agreement, together with any representations and warranties of Seller that survive the closing, are negotiated in the context of establishing a survival period, a "bucket" or minimum expenditure of Buyer, a "cap" or maximum liability of Seller, and whether the continuing obligation is supported by cash, a letter of credit or some other form of collateral security.

The survival period is intended to provide Buyer a sufficient period of time after the Closing Date of the transaction to become aware of facts and circumstances that might constitute a Seller breach of its representations and warranties under the Agreement. The appropriate period of time will vary depending on the nature of the transaction and the Hotel involved. For example, an Agreement covering the sale and purchase of a typical suburban Franchised Hotel may have a survival period of between six (6) and nine (9) months after the Closing Date, while an Agreement for the sale and purchase of a large managed Hotel with diverse amenities such as a resort or ski property may have a survival period of 12 or more months after the Closing Date. In hotel sale and purchase transactions when the ownership of the Hotel will change, but the management of the Hotel will remain in place, there is sometimes an exchange of letters or estoppel certificates between Owner and Manager that can help in this area, and other elements of the transaction as well. An Owner's Estoppel Certificate and Manager's Estoppel Certificate will set forth certain facts, such as the last date certain payments were made, and that neither party is aware of any defaults or facts that would mature into defaults upon the giving of notice or the passage of time, thereby supporting some of the representations and warranties in the Agreement.

In the event that Buyer believes it has a claim against Seller for breach of a representation, warranty or surviving indemnification, the claim must be asserted during the survival period, and the amount at issue must exceed some negotiated minimum threshold. Think of this as a form of deductible or self-insurance. Before Buyer can make a claim against Seller, Buyer's claim must exceed the minimum threshold. If the aggregate amount of the claims are below the threshold, the Buyer may not seek recourse to the Seller even though there is a breach of a representation, warranty or indemnity and it is within the survival period. Some Agreements will stipulate that once the amount of the claim exceeds the minimum threshold, the Buyer can make a claim against Seller for the full amount, including the amount under the threshold. The Buyer's claim reverts back to the first dollar spent by the Buyer as a result of the breach.

At that point, the next analysis is whether the Buyer's claim is subject to a maximum amount or "cap," regardless of the actual amount spent by Buyer as a result of the breach. This is a negotiated number that is often expressly or implicitly tied back to the purchase price, such as a negotiated sum representing an agreed upon percentage of the purchase price in the Agreement.

Then there is the question of collateral. The obligations of Seller are not always supported by collateral. This is a negotiation based on the nature of the Seller and the assets it continues to own after the Closing Date and the sale of the Hotel. More often than not, Seller will be a single purpose entity owning no assets other than the Hotel it has just sold. This has led to Buyers insisting on some form of collateral security for Seller's continuing post-closing obligations. This can take the form of a guaranty from Seller's parent entity or credit-worthy affiliate, cash held in escrow during the survival period, a letter of credit, some combination of all of these, or anything else Buyer and Seller agree to.

Working capital

On the Closing Date, the Hotel will have some amount of working capital in the Hotel accounts. The existing Management Agreement will also include working capital requirements and go further to provide procedures and time periods for the correction of any working capital shortfalls. The better practice seems to be that Seller would simply leave the working capital in the Hotel's accounts to be transferred to Buyer, often with a credit taken on the Settlement Statement, unless some other arrangement is negotiated between the parties. The working capital could have been factored into the purchase price. Branded Hotel Managers do not favor releasing working capital to then have to pursue a Buyer to put it back into the account, further supporting the concept of leaving the cash where it is, and handling any adjustments through credits on the Settlement Statement as between Seller and Buyer.

Checked baggage

A typical Agreement clause might look like this:

> On the Closing Date, representatives of Seller and Buyer shall make a written inventory of all baggage and similar items left in the care of Manager at the Hotel and all "lost and found" items belonging to guests (collectively, "Inventoried Baggage"). Buyer shall be responsible for, and shall indemnify, defend, and hold harmless Seller against, any liability, damage, loss, cost or expense incurred by Seller with respect to any theft, loss or damage to any Inventoried Baggage from and after the time of such inventory, and any other baggage or similar items left in the care of the Hotel on or after the Closing Date. Seller shall be responsible for, and shall indemnify, defend and hold harmless Buyer against, any liability, damage, loss, cost, or expense incurred by Buyer with respect to any theft, loss or damage to any Inventoried Baggage prior to the time of such inventory, and any other baggage or similar items alleged to have been left in the care of the Hotel on or prior to the Closing Date that was not inventoried.
>
> The foregoing notwithstanding, prior to the effectiveness of any obligation of the parties to indemnify one another under this Section, the party suffering loss shall exhaust available remedies under the Hotel Management Agreement due to such loss, if any, and, in the event Seller is the party suffering loss, Buyer shall reasonably cooperate with Seller's reasonable efforts to pursue its remedies, to the extent available. The indemnities set forth in this Section as well as the obligation to exhaust remedies shall survive Closing until the expiration of the Survival Period, and shall be subject to the limitations set forth in Section __.

Although this is a common type of clause in Hotel transactions, if there is professional management in place under a Management Agreement that will continue after the closing, then the tasks related to checked baggage are discharged by the Hotel Manager.

Safe deposit boxes

A typical Agreement clause might look like this:

> On or before the Closing Date, Seller shall cause Manager to notify all guests who are then using safe deposit boxes at the Hotel advising them of the pending change in the ownership of the Hotel and requesting them to conduct an inventory and verify the contents of such safe deposit boxes. All inventories by such guests shall be conducted by Manager under, to the extent practicable, the joint supervision of representatives of Seller and Buyer. At Closing, all safe deposit boxes which are then in use but not yet inventoried by the depositor shall be opened in the presence of Manager and, to the extent practicable, representatives of Seller and Buyer, and the contents thereof shall be inventoried. Following the inventory of each safe deposit box, Buyer shall indemnify, defend and hold harmless Seller against, any liability, damage, loss, cost, or expense incurred by Seller with respect to any theft, loss, or damage to the contents of any safe deposit box from and after the time such safe deposit box is inventoried. Seller shall be responsible for, and shall indemnify, defend, and hold harmless Buyer against, any liability, damage, loss, cost, or expense incurred by Buyer with respect to any theft, loss or damage to the contents of any safe deposit box prior to the time such safe deposit box is inventoried.
>
> The foregoing notwithstanding, prior to the effectiveness of any obligation of the parties to indemnify one another under this Section _____, the party suffering loss shall exhaust available remedies under the Hotel Management Agreement due to such loss, if any, and, in the event Seller is the party suffering loss, Buyer shall reasonably cooperate with Seller's reasonable efforts to pursue its remedies, including any claims covered by insurance, to the extent available. The indemnities set forth in this Section as well as the obligation to exhaust remedies shall survive Closing until the expiration of the Survival Period, and shall be subject to the limitations set forth in Section ____.

Here again, this is a common type of clause in Hotel transactions, that may be discharged by the Hotel Manager.

Insurance claims and related matters

The Purchase and Sale Agreement should address insurance claims and any other claims that might potentially be asserted against Seller both before and after Closing, together with claims that might be asserted against Buyer after Closing for which Seller might retain liability.

It is typical to find in the Purchase and Sale Agreement a schedule of pending litigation and claims, as well as the assignment and assumption of liability expressed in the agreement or in a separate assignment and assumption agreement. Language most often found in this context might simply state something as straightforward as Buyer expressly assumes the obligation for the performance of any and all of the obligations of Seller with respect to the Hotel and under the Management Agreement relating to the period on and after the Closing Date, but Buyer will not be liable for or have any responsibility for any acts or occurrences that may occur at or in connection with the Hotel or its operation prior to the Closing Date. Seller will retain liability for and have responsibility for acts or occurrences that arose or occurred at or in connection with the Hotel or its operation prior to the Closing Date. This concept may not be expressed precisely in this fashion, but there should be language addressing both claims covered by insurance and the action or inaction of Seller and Buyer to establish some date, usually the Closing Date, as the dividing line between obligations and liabilities assumed by Buyer and obligations and liabilities retained by Seller.

To fully express all of the elements of this concept of assumed or retained liabilities, most Purchase and Sale Agreements will connect this to the survival period and the negotiation of a funded escrow or other collateral security to support Seller's covenants. These were explained earlier in this chapter.

3 Hotel Management Agreements

General considerations of the Letter of Intent

The Hotel Management Agreement is the central governing document for the relationship between the Hotel Owner and the Hotel Manager. While the Hotel Owner is the party that bought or built the Hotel, financed the Hotel, and may have even provided a personal guaranty of all or a portion of the debt secured by the Hotel, it is the Hotel Manager that is responsible for the day-to-day operations of the Hotel and the task of achieving compliance with Brand Standards, maintaining compliance with all applicable laws, making the Hotel excellent from the perspective of the guest, as well as profitable from the perspective of the investor or Mortgagee.

A letter of intent (LOI), term sheet, memorandum of understanding ("MOU"), or similar type of document typically is the initial document negotiated by the Hotel Owner and Hotel Manager. The LOI is traditionally a non-binding agreement, but for certain components of the LOI such as the agreement of the parties to deal exclusively with one another for some period of time, and agreeing to maintain and preserve as proprietary and confidential some of the information that the parties may share during this preliminary period of exploration. One school of thought holds that the LOI should be brief and describe primarily the basic business terms of the relationship, and leave for the negotiation of the Management Agreement the majority of matters frequently negotiated by the parties. This can lead to some vague or loosely drafted language finding its way into the Management Agreement. By way of contrast, there is another school of thought that advocates using the LOI discussions to flush out many of the other issues that each side knows is coming, and use the LOI, despite the fact that it is not binding, to negotiate in advance some of these other more difficult issues.

Proponents of the early discovery and negotiation of issues will consider all of the following elements of a Management Agreement as potentially worthy of early negotiation in the LOI.

The LOI should describe the scope of the proposed project to be undertaken by Owner and Manager. Independent third-party hotel managers offering an Owner an LOI for a Hotel to be operated pursuant to a Franchise Agreement or as an independent hotel may have an LOI that is a little less dense than a Branded Manager. Nevertheless, both will use the LOI in the same manner and either avoid issues or flush out issues, depending on which school of thought is being adhered to. This may include the following:

- the Hotel and any associated residential project;
- the Hotel Management Agreement for the operation of the Hotel;

- the Brand standards of the particular Brand to be associated with the project;
- the Technical Services Agreement ("TSA") by which Manager will lend expertise to Owner's design and construction consultants in an effort to assure conformity with the design and construction standards that will apply to the development of the Hotel;
- the pre-opening services agreement describing the functions of Manager during the pre-opening period;
- fees and agreements for Centralized Services;
- if the project will include real estate that will be sold by Owner, the residences sales and marketing license agreement by which Manager will license its "Brand" and "marks" to Owner in connection with the sale and marketing of the residences;
- if the project will include real estate that will be sold by Owner, a Residences Management Agreement ("RMA") pursuant to which Manager will manage and provide services to the residences;
- the governance regime, such as a condominium form of ownership and various covenants and declarations.

These functions and relationships can be memorialized in separate agreements, but that is not necessarily the case. Oftentimes some of these agreements are combined, so it should not be assumed that every document will be in every transaction.

In the context of a Management Agreement with a "Brand," to the extent the information is sufficiently developed and available, such as when a detailed PIP has already been prepared, the LOI should confirm the agreed-upon deviations from Manager's "Brand Standards" or as it relates to a rebranding or PIP, for design and construction, such as, room size, room design, public areas, bathroom elements and fixtures, type/size of dining facilities, and recreation and spa improvements. When less information or less detailed information is available, it may simply be impractical for the LOI to address the potential issues surrounding the PIP, but any steps that can be taken at the stage of the LOI to draw out potential issues will save time and energy when the Management Agreement is being negotiated. For example, if the Brand standard includes a swimming pool but the proposed site cannot accommodate a pool, that deviation must be memorialized and approved by the Brand. If schematic drawings are available, it serves the interests of both Owner and Manager to commit these deviations from "Brand Standards" to writing based on the preliminary or schematic design drawings.

The LOI should reflect the level of services and the fees for those services to be rendered to any residential real estate that may be sold to third parties as Branded residences, including the segregation of those services that will be deemed mandatory and those services that will be deemed optional to be provided on an "à la carte" basis. This portion of the LOI might specify that the Hotel portion of the project would provide a platform for Manager to offer to the residential Owners services such as concierge services, laundry services, in-room dining and a variety of Hotel amenities. These provisions will ultimately be embodied in the RMA.

An early discussion of a restricted territory or area of protection is important. Owners will seek an exclusivity or restricted area provision preventing Manager that is a Brand from owning, licensing, franchising, or managing a "competing Hotel" within the radius of the exclusivity or the territorial area. These types of provisions come in various shapes and sizes, and can be structured to be maintained during the entire term of the Hotel Management Agreement. The definition of what is a "competing Hotel," to not only include the Brand selected, but any other Brand of the Branded Manager's family of Brands (an existing Brand or one that is subsequently introduced) which has the potential of competing against Owner's Hotel can be discussed but is generally unavailable to Owners. It is exceedingly rare to see in transactions currently being negotiated today an area of protection that crosses Brands and is other than restricted to the single Brand of the subject Hotel. Moreover, the Branded Manager's grant of a restricted territory to Owner will be subject to Manager or any of its Affiliates' acquisition, whether through purchase, sale, merger, consolidation, or other type of transaction, of another chain of hotels, franchise system, group or portfolio of some negotiated number of hotels, or the acquisition of the right to operate or manage another chain, franchise system, group or portfolio of that pre-negotiated number of hotels. The impact of this is that if there is a hotel, or even more than a single hotel located within the restricted territory, Manager and/or its Affiliates have the right to convert a hotel, and although it should be negotiated and specified, it is generally just one hotel, of the newly acquired hotels to the Brand of the subject Hotel, which can then be operated under the same Brand as the subject Hotel. From Manager's perspective, the negotiated number of hotels necessary to constitute a chain, system or group of Hotels should be small, such as five hotels, and very often under ten hotels. From Owner's perspective, since the effect of the transaction constituting a chain of Hotels is to permit Manager to operate another Hotel under the same Brand as the subject Hotel within the restricted territory, the pre-negotiated number should be large, such as not fewer than ten hotels.

With so many Owners being driven by a relatively short "hold" period for the Hotel, as compared with the relatively long term of a Hotel Management Agreement, Owners and Managers may have no alternative but to negotiate Owner's right to sell the Hotel and terminate the Hotel Management Agreement by offering to make a termination payment to Manager consisting of a multiple of prior years' management fees or some other negotiated formula. This historically has been, and may continue to be, very challenging for some Managers, but it is a conversation that should occur, particularly in the context of a Management Agreement with Third-Party Manager rather than a Brand Manager.

International Projects will present additional concerns and sensitivities that might be addressed early in the LOI. In addition to all of the issues surrounding a Hotel project in the U.S., there are many other issues that arise in the context of projects developed outside the U.S. We will briefly mention a few of those issues, but often the drive to address these issues in the most advantageous manner possible will result in a system and structure of documentation that is very different from that of a domestic transaction. The matters to be addressed in the definitive documents and the dynamic tension between Manager and Owner will be similar and often the same, but the structure of the relationship may be different. One simple example is that the management fees may be allocated in a different manner and arise in a variety of documents not seen in domestic transactions, such as an International Services Agreement, Onshore Management Agreement, Offshore Management Agreement and Licensing and Royalty Agreement.

Tax Withholding and "Gross-Ups" can be important modifiers of management fees. Some Managers readily absorb the taxes imposed on their fee income derived from abroad, while others insist that Owner absorb Manager's local income tax by having the Hotel operation or Owner pay the required withholding taxes on the remittances of Manager's fees, coupled with a "gross-up" of such taxes so that Manager receives its fees net of the local income tax impact.

International management transactions require an early discussion of who will employ the employees. Manager will often require Owner to be the employer of the personnel assigned to the Hotel. In these circumstances, care is required to ensure that Owner is adequately indemnified for Manager's gross negligence or willful misconduct in its management, hiring, supervision and termination of Hotel employees. There is also a special need for the parties to allocate responsibility for "ex-pat" relocation expenses and the cost of severance and retirement benefits that may have accrued prior to any employee's tenure at the Hotel. To the extent permitted under local law, Manager and Owner may explore creating an "Employer Entity" that is affiliated with either Owner or Manager to act as the employer. This does not alter Manager's role to hire, supervise, train and discharge employees, but it will shift some burdens and benefits away from Manager and Owner.

Attending to the LOI with a heightened level of care and attention is being driven by the tendency to treat the LOI as non-binding when one of the parties does not wish to proceed beyond the LOI to the definitive agreements, but to treat the LOI as binding when the parties move on to the definitive agreements. The LOI then establishes the baseline for the coming negotiations and material deviations are resisted by both parties as "re-trades."

Key definitive agreements and terms

After the LOI has been negotiated and signed, the parties will commence negotiation of the definitive agreements. In recognition of the value of the management arrangement to all Hotel Managers, of every type, style, and kind; Branded Manager, independent third-party manager, large international management company or small local management company, it is Manager that will begin the process and produce the initial forms of all operative documents. Manager will deliver to Owner the form of Hotel Management Agreement that Manager has developed and refined over time based on Manager's cumulative experiences with many Owners and any relevant judicial decisions and laws. It is from this point of beginning that the Hotel Owner will be negotiating. As may have already become clear, it is a simple fact of hotel legal practice today that a mature, well developed and robust Hotel Management Company will begin the process with its documents and the preparation of the definitive agreements will proceed from there. The Hotel Owner will be reacting to what the Hotel Manager has drafted into its standard form documents, and not the other way around. To make certain that all matters are covered, all of the definitive documents should be concluded before any one of them is signed. A brief summary of the key agreements is set forth in the following.

Hotel Management Agreement

The negotiation of the Hotel Management Agreement will be the most time consuming and controversial because of the long-term nature of the agreement and the sheer magnitude of the document. The Hotel Management Agreement is a long-term agreement that can have an initial term of between five and 30 years. The initial term may be followed by a series of renewal terms, creating an aggregate term that can exceed 50 years, particularly in the luxury segment. The proposed Manager of the Hotel, be it a well-known international Branded Manager or an independent third-party Hotel Manager, will present its form of Management Agreement to Owner.

The independent third-party Hotel Manager will generally have a form of Management Agreement that is shorter in length and with terms and conditions that Owner may have greater ease negotiating than the Brands, but, regardless of who the Hotel Manager is, Owner and Owner's advisors will start by examining and negotiating a document that Manager considers to be Manager's tried and true and most favorable form of document. This also means that Owner is not only confirming consistency with the LOI and working to understand the approach of Manager to Hotel operations, but also trying to identify things that are absent from the Hotel Management Agreement. It can be challenging for less experienced Owners with less experienced advisors because they may be unaware of what might be added to a Hotel Management Agreement to benefit an Owner if Owner would simply ask. Although each form of Hotel Management Agreement is unique, there are some typical significant issues such as the following.

Term

Manager will suggest a long-term contract. Owner will want to avoid being bound to a long-term Hotel Management Agreement, although in the context of a Branded relationship achieving a shorter term will be challenging. When the project is not just a Hotel, but includes various mixed-use components including Branded residential real estate, part of the appeal in selling the residences will be the benefits of long-term Branding by a major Hotel company and the services of the adjacent Hotel that operates under the same Brand. One solution is a significant initial term followed by a series of renewal periods based on the mutual agreement of Owner and Manager. Stronger Managers will seek to secure multiple, automatic renewal terms, but stronger Owners will make this subject to achieving an Owner's Priority Return or other performance metric(s). The concept of achieving a return to Owner as a condition for renewal of the term of the Management Agreement is not the same condition or test as the Performance Test to be examined later. The Performance Test is a metric that gives rise to an Owner's right to terminate the Management Agreement, and if the Performance Test is also tied to an Owner's Priority Return it is likely to include different thresholds because it is seeking to achieve a very different result.

Management fees

The LOI should specify the financial terms of the Hotel Management Agreement, including the description of the Base Management Fee ("Base Fee"), incentive management fee ("IMF"), fees for other services (such as IT, accounting, local marketing, and Brand advertising), the technical services fee under the TSA, the pre-opening services fees and charges, reservation services fees, and, if there is a Branded residential component, a fee based on a percentage of the gross sales price of the residential unit or a fee based on square footage of a unit, if a price per square foot is achieved above a base line, as being attributable to the power of the Brand. The fundamental underlying economics for the relationship between Owner and Manager is best addressed and resolved in the LOI and not deferred to the definitive agreements.

Base Fees have customarily fallen within a market rate percentage of GOR. The Base Fees usually "ramp up" during the period following the Hotel's opening date that is intended as a negotiated Stabilization Period. An IMF is intended to motivate Manager to be mindful of the expenses it incurs on Owner's behalf, and better manage the cost of generating revenue. Because the Base Fee is a percentage of the Hotel's top line gross revenue, Manger will earn its Base Fee without regard to the operating expenses and fixed charges of the Hotel. The IMF, on the other hand, is earned through Manager's efforts to make a profit at the Hotel. For Manager to truly prosper, Manager must first operate the Hotel so that Owner will prosper.

The calculation of the IMF can take many forms and the only limitation will be the limits established by Owner and Manager in the negotiation of any particular transaction. Nevertheless, a few examples are worth describing in more detail. One such example is an IMF calculated as a percentage of "Operating Cash Flow." Operating Cash Flow is customarily expressed as "Adjusted Net Operating Income" on an accrual basis, excluding interest income and expense, depreciation and amortization, but after deducting (a) the amount contributed to the annual furniture, fixtures and equipment reserve ("FF&E Reserve"); and (b) an agreed upon return ("Owner's Priority") on "Owner's Total Investment." The calculation of the IMF may also be calculated as a percentage of Adjusted Gross Operating Profit ("AGOP") above a threshold percentage of gross operating profit ("GOP"). AGOP is equal to GOP minus the Base Fee. The financial and accounting terms used in Hotel Management Agreements are typically defined by and determined in accordance with the Uniform System of Accounts for the Lodging Industry. The so-called "Uniform System" was initially published in 1926 by the Hotel Association of New York City to establish a uniform accounting system for the lodging industry. The Uniform System has been, and continues to be modified to provide an up-to-date accounting methodology for the industry, and remains the primary basis for the calculation of financial data in the Hotel industry. Because the calculation of the IMF will flow from other financial calculations as defined pursuant to the Uniform System, Owner and Manager will negotiate mutually agreeable components of each defined term unless they agree to accept the Uniform System definitions without modification and because there are so many variations to the calculation of the IMF. (Please see the companion website and visit www.routledge.com/cw/migdal to view the examples. Because there are so many defined terms that may apply, refer frequently to the *Glossary*.)

Owner's Total Investment is intended to capture all sums expended by Owner to acquire, develop, construct and finance the Hotel, including hard and soft costs and loan proceeds. It might also capture all additional sums expended for capital improvements beyond the sums funded into the FF&E Reserve after the Hotel opens. The definition of Owner's Total Investment should include the acquisition and development loan proceeds amount and actual construction period debt service (or debt service based on a predetermined "base constant" rate) during the construction period. If the capitalization of the project contemplates mezzanine financing, Owner's Total Investment should include the amount of mezzanine loan proceeds and construction period mezzanine loan interest.

If a GOP percentage is used to arrive at the IMF, Owner's Total Investment is irrelevant, and the focus becomes the percentage of GOP below which Manager would not earn an IMF. In this situation, Manager must cross a negotiated threshold of GOP before earning the first dollar of an IMF. It is often considered an effective vehicle to better align the interests of Manager and Owner to have Manager work to earn more than simply a nominal GOP in order to begin earning its IMF.

Another sensitivity in the marketplace today is to see Owners negotiate to limit the IMF to a percentage of true "bottom line" revenue to Owner, that is, cash available for distribution after payment of all expenses, including fixed charges (other than post-opening debt service), and Owner's Priority. In calculating the IMF, key issues include whether there will be a cap on Owner's Total Investment, what items will be included in Owner's Total Investment (such as initial and future working capital), and the amount of future major capital expenditures that will be included (usually sums deposited annually to an FF&E Reserve are deducted in arriving at Operating Cash Flow, while capex beyond such reserve amounts is added to Owner's Total Investment). Commonly, the calculation of Operating Cash Flow and, therefore IMF, is limited to an annual rather than cumulative basis. It is worth noting, however, that the past practice of many Managers of using a GOP percentage to calculate the IMF is reemerging. Managers are insisting that they cannot control and therefore should not be penalized by escalating fixed charges, such as taxes, insurance and rental costs. Owners take a calculated risk in moving from an Operating Cash Flow IMF calculation to one based on GOP margins. If a GOP percentage is used to calculate the IMF, some Owners insist that no IMF be paid unless GOP exceeds a specified percentage threshold of GOR. In this manner, Owners should be able to cover the costs of the fixed charges and may be able to assure a return equivalent to Owner's Priority used in calculating the IMF as a percentage of Operating Cash Flow. As a practical concern and strategy of an Owner, Owners should delay any agreement on the IMF calculation until a market feasibility study and at least a five-year financial proforma for the project has been developed and analyzed. It is always a good practice to obtain Manager's proforma and have it verified with a second proforma prepared by an independent consultant.

Identity of services and fees for other services

The Hotel Management Agreement will not only specify Manager's Base Fee and IMF, but Manager will also charge a variety of fees for their customary program and Brand services, such as sales and marketing, central reservation systems, accounting fees, guest reward programs and technology fees. A valuable strategy is for Owner to obtain a full description of the scope of services and a schedule of all fees for these services prior to signing the LOI, along with a proforma reflecting the Hotel's performance for at least the first five years of operations, detailing all revenue and expenses, including these fees. A savvy Owner will also request a copy of Manager's market feasibility analysis report for the project, although many Managers have ceased furnishing their feasibility reports to Owners. In the absence of receiving a feasibility study from Manager, Owner should engage an independent hospitality consultant to conduct such a study. Caution will need to be exercised with respect to long-term proformas. Managers are able to predict the future no more accurately than Owners, so any proforma, regardless of who prepares it, will be no more than a good faith effort to predict future performance based on then existing baselines, without warranty or representation that the proforma will be achieved and without liability on Manager for failing to achieve the predicted results.

Termination-on-sale

Owners will want to secure the right to terminate the Hotel Management Agreement upon a sale of the Hotel. In the eyes of some, the sale of a Hotel "encumbered" by a Management Agreement will diminish the price and limit the universe of potential buyers, including other Branded Hotel Operators. In the eyes of others, the sale of a Hotel "enhanced" by a Management Agreement with a professional Hotel Manager will only increase the price and the desirability of the asset. Although some Managers may initially vigorously resist a termination-on-sale right, the solution is often achieved by agreeing to a "lockout" period during which the Hotel Management Agreement cannot be terminated on sale, and then at the end of the lockout period, a sale of the Hotel is permitted. To their credit, many Hotel Managers, including the large Brands with deep commitments to the long-term nature of their agreements because it directly impacts the value of the company, recognize that the Hotel Owner of today, particularly when the Hotel is of significant size or stature, is likely to include parties with a limited horizon for holding the Hotel and a clear exit strategy that will require execution long before the natural end of the initial term of the Management Agreement. This has led to better solutions where the sale of the Hotel can be accomplished, but the third-party purchaser has to meet certain clearly described criteria, including not being a competitor of the Hotel Manager. Beginning at the end of a lockout period Owner may terminate the Hotel Management Agreement upon payment of a "termination fee" equal to a multiple of prior years' management fees (a combination of Base Fees and IMF). One good strategy is to let the termination fee multiple decline as the years elapse (e.g., a three-times multiple if terminated in operating years eight through 11, a two-times multiple if terminated in years 12 through 15, etc.). The next best exit solution to sale termination rights is the right to convert the Hotel Management Agreement to a Franchise agreement in connection with the sale of the Hotel. Because Hotels that operate in higher quality Brand segments typically are not Franchised, this exit solution may not be available for a luxury Brand Hotel. However, on the theory that a Brand that currently does not Franchise may do so later, an Owner should nevertheless consider securing a Franchise conversion right which may be triggered as soon as the particular Brand is subject to Franchise arrangements. The termination fee that an Owner might negotiate in connection with a sale of the Hotel to an unaffiliated third-party may be very different from provisions in a Management Agreement that specify liquidated damages in the event of an Owner's wrongful termination of the Management Agreement. Owners, particularly institutional Owners that may acquire a Hotel with an exit strategy and exit horizon in mind from the outset, will generally make a decision to sell the Hotel for any number of legitimate business reasons that are completely unrelated to its operating success and the relationship with Manager. If the Hotel buyer and potential new owner is a permitted and qualified person under the Management Agreement, but does not desire to assume the existing Management Agreement, Owner may still desire to pursue the sale of the Hotel to that buyer. The negotiated termination fee permits Owner to proceed with the sale, without the continuation of the Management Agreement, but with compensation to Manager for its loss of the Hotel, and the fees Manager would have had the ability to earn under the Management Agreement. Liquidated Damages, in the context of the termination of the Management Agreement when not permitted by the agreement, offer a completely different set of dynamics. At present there is a lack of consensus on the efficacy of pre-negotiated liquidated damages provisions in Hotel Management Agreements. The inclusion of such a provision can save significant amounts of legal and professional fees and time by avoiding adversarial proceedings just on the appropriate measure of damages once liability has been established. If Owner and Manager can agree to include a liquidated damages provision in the Management Agreement and then further agree upon the proper formula or methodology for the determination of the liquidated damages, then further care will need to be exercised in the drafting of those provisions to establish the foundation for the liquidated damages as a fair and reasonable manner in which to assess damages that are not otherwise determinable and, at the same time, not a penalty to the party paying the damages.

Performance test termination rights

A fulsome and complete LOI will include a meaningful performance test termination right for Owner. Customarily, these tests are not applicable for a period of time after the Opening Date of the Hotel. The assumption is that this period is necessary for the Hotel's operation to be stabilized and the market position established. Once this period, often defined as the Stabilization Period, expires, the Performance Test becomes applicable. The duration of the Stabilization Period will be negotiated on a case-by-case basis and is heavily dependent on the Hotel itself and the market in which it is located. The Performance Test is generally based on the Hotel's failure to achieve *both* a percentage of a pre-approved budgeted GOP *and* a percentage of revenue per available room or "RevPAR" for a defined "competitive set." Some Owners will try to negotiate a stand-alone "minimum return test" requiring Manager to deliver the agreed on minimum return, failing which, for two consecutive fiscal years, Manager can be terminated without any termination fee payment by Owner or claim by Manager. This is particularly useful in a market area that lacks a representative competitive set for RevPAR test purposes, such as resorts in remote destinations, or markets that are not fully developed or markets that do not report to Smith Travel Research. This is different than Owner's Priority amount or "hurdle" for purposes of calculating the IMF discussed earlier. This element of the LOI is not directed toward when Manager might earn an IMF, but rather toward Owner's ability to terminate Manager without payment or a termination fee based on performance. Manager will resist a "minimum return test" termination right in favor of Owner arguing that, as a Manager (and not an Owner), it should not have to take any "market risks." Owner, by the same token, will want to secure the right to terminate the Hotel Management Agreement if the economic performance of the Hotel is unsatisfactory. The solution is often a two-pronged performance test that is based on two consecutive years' performance commencing after the negotiated Stabilization Period. The test that Manager will offer usually involves its failure to achieve a percentage of the agreed upon budgeted GOP ("GOP Test") *and* its failure to achieve a percentage of the revenue per available room ("RevPAR") generated by the Hotels identified as the "competitive set" ("RevPAR Test"). Care should be taken to properly select the competitive set of Hotels. Once selected, there should also be a process to update and modify the Competitive Set over time, particularly in light of the long-term nature of the Management Agreement. For example, some Management Agreements take this style of approach:

> If the competitive set requires adjustment due to changes in the market or in the hotels then included in the Competitive Set, then either Owner or Manager may suggest an alternative Competitive Set for the other Party's Approval during the Annual Business Plan approval process. If the Parties fail to agree on an alternate Competitive Set within a reasonable period of time, then at the election of the suggesting Party, the matter shall be resolved by the Expert Resolution Process.

There has been criticism of the GOP Test as one prong of the two-pronged test described above because the budget is negotiated by Owner and Manager each year, thus allowing Manager to influence or limit the target GOP. Managers negotiate for the performance test to be suspended in an operating year during which there is a budget impasse. Consequently, stronger Owners often negotiate for an independent "minimum return test" permitting termination of the Hotel Management Agreement if Owner has not received a negotiated percentage of Owner's Priority.

Even when there has been a Performance Test failure, many such tests include a right to cure in favor of Manager through paying Owner, for example, the shortfall under the GOP or Minimum Required Return Test for the second failed year. A stronger Owner may successfully negotiate for repayment of the shortfall for both years, reasoning that the failure of the test required two consecutive years of failure. The LOI should commence this conversation early, because Manager's cure right will generally lead to another negotiation of how many times Manager may exercise the cure right. Managers will seek a continuing or "evergreen" right to make a shortfall payment and cure a performance failure. There will also be language to render a performance failure inapplicable in the presence of other factors outside the control of Manager, such as force majeure events, Owner failing to perform under the Management Agreement, and a number of hotel rooms being removed from hotel inventory. Owners will assert that when a Manager fails the Performance Test year after year it is indicative of other problems and there has to be a limit on the number of cure payments that Manager can make. There are important considerations on both sides, which is why this matter should be negotiated early.

Non-disturbance

Although the Hotel Management Agreement does not, technically, grant an interest in real property, Managers will endeavor to obtain the agreement of Owner that Owner's Mortgagees will not disturb Manager or seek to terminate the Hotel Management Agreement in the event the mortgage is foreclosed (absent Manager's uncured default). In market-challenged times Owners fare better in resisting a Manager's request for a non-disturbance agreement from Mortgagees. More recently, however, with the return of some flexibility in the capital markets for Hotel properties, it is not uncommon for a Hotel Owner to agree to exercise "commercially reasonable efforts" to obtain a non-disturbance agreement from the Mortgagees in favor of Manager. However, it can be a challenge for an Owner to provide within the Hotel Management Agreement an absolute commitment to provide a non-disturbance agreement. An Owner would be well served to indicate that Manager may not receive a non-disturbance agreement from the Mortgagees. This is often a serious point of negotiation for Managers. A common (but often unsatisfactory to everyone) compromise is to resist a non-disturbance agreement, but agree to limit mortgage financing to a maximum loan-to-value test or loan-to-cost test, and/or a minimum debt service coverage ratio requirement. An alternative approach is requiring non-disturbance only if mortgage debt exceeds a negotiated percentage, as measured with each financing, but not below a negotiated percentage LTV or LTC leverage, and requiring "Recognition Agreements" from all mezzanine Mortgagees (i.e., the equivalent of a non-disturbance agreement). Care should be taken not to agree to a maximum mortgage and mezzanine debt level, but, increasingly, Managers are insisting on leverage limitations to avoid the market tarnish of a foreclosure or bankruptcy of one of their Branded Hotels. Many Managers will take the position that non-disturbance is non-negotiable at any leverage percentage, and will simply require an Owner to obtain non-disturbance from the Mortgagee. Non-disturbance and comfort letters are examined in deeper fashion in Chapter 6.

Transfer rights

In connection with Owner's sale of the Hotel, Manager will typically want the right to approve the new Owner as well as a right of first refusal or right of first opportunity to negotiate to acquire the Hotel with respect to the proposed sale. The right of first refusal can be very challenging for an Owner, and thought should be given to the approach of limiting any purchase rights in favor of Manager to a right of first offer or, better yet, a limited right of first negotiation. Manager's purchase opportunity should be structured in such a manner as to provide to Manager the limited opportunity to preserve an asset that might be of strategic importance, but not have a chilling effect on Owner's opportunity to sell the Hotel for a favorable price. This is often accomplished through a short-term right to negotiate the basic terms for a sale to Manager. Even if Manager is not given purchase rights of any kind, it will seek to prohibit a sale to an undercapitalized buyer, a felon, or a Competitor. The restriction of a sale to a competitor is problematic in today's environment due to the blurred distinctions between Branded Hotel companies, which are competitors, and companies that start off as Hotel funds or Owners, but later acquire Branded identities. A Hotel company that is not a competitor today could become a competitor tomorrow and vice versa. Therefore, Owner should be careful to try to negotiate language that eliminates the "no-competitor" restriction. Alternatively, Owner should try to limit "competitors" to Hotel companies with (i) an established Brand identity, (ii) a system-wide centralized reservation system, and (iii) of a chain scale (e.g., upscale, upper-upscale, luxury, etc.) that is truly competitive with Manager. Manager's concerns about the potential purchaser being a Competitor are legitimate and should be fairly addressed. The Owner–Manager relationship requires a high degree of communication and collaboration and this will be impaired if the potential purchaser competes with Manager in the hospitality business.

Radius restriction or area of protection restriction

Owner should negotiate Manager's agreement not to own, operate, Franchise or otherwise have an interest in another Hotel within the defined territory for as long as possible, if not for the life of the Hotel Management Agreement. Manager will resist radius restrictions altogether and, secondarily, seek to limit its application to Hotels operating only with the same Brand and for only a portion of the term, meaning that Manager may insist that the restricted territory shrink and eventually "burn off" over time. At a minimum, Manager may insist on a "chain acquisition" exception to the radius restriction, meaning that it may acquire a chain of Hotels (the higher the number required to constitute a chain of Hotels the better for Owner) of the same chain scale (e.g., upper-upscale or luxury). This may be acceptable provided Manager agrees not to re-Brand any of the Hotels acquired in the chain acquisition (or merger) to the same Brand as Owner's Hotel. Although these exceptions are commonly granted, Owners remain compromised even if, following a chain acquisition, Manager does not re-Brand a comparable class of Hotel, once the two comparable assets are linked by a common centralized reservation system, the goal of the radius restriction can be defeated. Another more recent trend that challenges the impact of an area restriction is that Managers have and will continue to create "Brand derivatives" that are not bound by the radius restrictions. Under a Brand derivative exception to the radius restriction, a similar Brand scale Hotel that includes the Brand name as part of its name would be free of the restriction (e.g., "XYZ Brand" vs. "the ABC, an XYZ Collection Hotel"). A further challenge is to avoid the radius restriction by differentiating Hotels of the same Brand based on "category" (e.g., resort, convention, airport); but the distinctions among the various types of "category" Hotels are so miniscule that, in practice, they will likely represent a competing Hotel within the same restricted area. A very strong Owner may attempt to negotiate into the Management Agreement a right of termination if Manager becomes involved in another hotel, as owner, manager, licensor or many other ways, and that other hotel, although permitted within the Restricted Territory under the Management Agreement, is or becomes competitive with the Hotel and causes a materially adverse financial effect on the first Hotel. The challenge here is one of measurement. Without a reasonably objective criterion through which to monitor and measure the impact of the second hotel on the first hotel, that includes a methodology to account for the differences between the two hotels, it is going to be very difficult for a Hotel Manager to agree to permit an Owner to terminate the Management Agreement in this situation. The conversation between Manager and Owner may become analogous to the discussion of an area of protection in the context of a Franchise Agreement. Many Franchisors have employed the use of an impact study to calculate the effect of a new hotel on an existing hotel, which may provide the information needed to assist in the resolution of this type of issue.

Budgets

The first draft of the Hotel Management Agreement will typically grant Manager almost absolute authority over day-to-day operational matters at the Hotel, such as setting room rates, hiring and firing employees, determining salary, and sales and marketing efforts. Consequently, Owner's budget review and approval rights are critical to controlling costs. Owner should fastidiously review and negotiate each annual operating and capital expenditure budget and limit Manager's authority to overrun expenses by aggregate and line item maximum increases (typically in the range of five to 10 percent), with exceptions only for costs that increase with volume, because with increased occupancy comes more demand for supplies and maid service, and for uncontrollable expenses such as insurance, real estate taxes, weather-related costs, and unforeseen increases in the cost of utilities.

Cash management

Somewhat related to the non–disturbance issue, cash management can be a contested topic. The Mortgagee will require a perfected security interest in all cash generated by the Hotel. This is customarily achieved by granting language in the Mortgage, together with a "Controlled Account Agreement," over each and every account of the Hotel. Mortgagees customarily demand that all sums generated by the Hotel be funneled through a "lockbox" and cash management account, with subsequent remittance to Manager for the payment of operating expenses, payroll and management fees. Managers, on the other hand, will vigorously resist any cash management regime that does not permit Manager to freely pay operating expenses and its Base Fees and reimbursable expenses in the ordinary course of business, irrespective of a mortgage loan default. Many Mortgagees will insist that following the occurrence of an event of default, the Mortgagee should control all of the cash and be permitted to apply the cash to debt service, even prior to the payment of operating expenses and Base Fees. This is particularly true for loans advanced during credit restrictive times and for loans that are rated by a credit agency. The nature of a hotel as an operating business that is always open for business, and usually subject to professional management, should help both Owners and Managers explain to Mortgagees not typically active in the hotel industry, that the typical non-hotel lockbox is not always a workable mechanism and needs to be re-examined. A common compromise is to give the Mortgagee full control over all cash and all accounts of the Hotel (by controlled account agreements on each account and a cash Management Agreement), but to permit Manager access to the accounts in order to pay budgeted operating expenses and its Base Fees and reimbursable expenses until such time as the Mortgagee elects to accelerate the loan following a mortgage loan default. All other sums, both pre- and post-loan acceleration, would be subject to the cash collateral account, but following loan acceleration, the Mortgagee could elect to pay debt service before operating expenses and management fees. The larger, more established Hotel Managers often succeed in persuading Mortgagees to permit them to pay operating expenses in the ordinary course even after an event of default under the mortgage, and empirical data suggests this is a sound long-term decision; Hotels that are cash starved often suffer more rapid decline in down cycles than those that are properly sustained and competently managed through a cash trap or foreclosure period.

Technical Services Agreement

The Technical Services Agreement or "TSA" is a document by which Manager will deliver its Brand supervision services during the design plan preparation and construction stage. Specifically, Manager will use the TSA as a means of reviewing and approving all of the design plans for the Hotel and, to a lesser extent, the construction plans, to assure compliance with Manager's design criteria. In addition to negotiating the fees for such services, which may be an agreed upon "flat fee" or a price per Hotel room or "key," the TSA will govern the protocol for Owner's submission of all design and construction plans to Manager, the duration and scope of Manager's review rights, and the schedule for the design process and periodic meetings between Owner and its consultants and Manager's representatives. Sometimes the TSA fees do not rationally correlate to Hotel room count and, instead, are "capped" at a fixed fee. This is particularly appropriate in mixed-use projects where Manager is reviewing design elements of residences and large common spaces and amenities improvements, in addition to the Hotel design.

Although appearing to be an ancillary document, the TSA is of critical importance to both Owner and Manager. Both parties benefit from the careful negotiation of rights in the TSA committing Manager to periodic review and approval of the plans at each stage, and to prevent Owner and Manager from changing and renegotiating an element of the plan that has been previously approved. At each point in the planning process, the parties should carefully document any agreed on deviations from Manager's Brand Standards. This should begin at the very early schematic plan stage. There may arise situations, which will be specific to the project, when it is beneficial to deviate from the general guideline of executing the TSA and the Management Agreement together, and execute a TSA or letter agreement reflecting certain general early stage planning and design provisions in order to not impede the pace of the overall development process. The TSA should commit both parties to submit or review a submittal within a certain period of time, failing which, the submittal would be deemed approved. Matters that have been approved by Manager or deemed approved cannot be withdrawn or subject to changes in the Brand Standards after approval. The potential expense and time lost to Owner could be devastating.

In the event of a disagreement regarding the application of the Brand standards, senior management of Owner and Manager should meet for the purpose of a one-day mediation to resolve the matter. This should cause Manager's design representatives to be cautious in their insistence on Brand Standards that are inapplicable or impractical due to market conditions or geography. Failing resolution of the dispute through a non-binding mediation with senior management, either party could then avail itself of arbitration with competent design and legal professionals or experienced hotel advisors acting as the arbitrators. Arbitration sponsored by AAA is a common forum for resolving these types of disputes.

During the design, development and construction stage, Manager and Owner will debate and discuss whether there should be hard milestones, the failure of which would require payment of increased technical services fees to Manager or permit Manager to terminate the Management Agreement. These milestones might include dates for obtaining a building permit, closing on a construction loan, commencing construction of the Hotel, completing construction of the Hotel and opening the Hotel for business. The TSA should also provide a list of Owner's pre-approved project consultants, including architects, planners, civil engineers, and interior designers or a mechanism for selecting them from a list of recommended or approved professionals provided by Manager. Although Manager's representatives should be involved on a weekly basis, Owner will often seek to limit Manager's formal review and approval rights of plans to certain defined stages of the design development process, such as the schematic drawings, design development drawings, and final plans and specifications. If, despite liberal milestones, Manager is able to terminate its relationship with the project, Owner and Manager might then be negotiating whether Manager can "revive" the agreement in the event that Owner subsequently reinitiates the project.

Technical Services are examined in deeper fashion in Chapter 7.

Pre-Opening Services Agreement

The Pre-Opening Services Agreement, whether separate or contained within the Hotel Management Agreement or TSA, is the document by which Manager will identify its operating standards, usually by reference to a published operating manual, which may be made available through a password protected website. In anticipation of the opening of the Hotel, the Pre-Opening Services Agreement will identify all pre-opening expenses for the pre-opening activities. The expenses are customarily identified through a detailed pre-opening budget that is attached to the agreement. The pre-opening services customarily involve the following:

- the sales promotion of the Hotel;
- preparation of programs to secure business and bookings for the Hotel's facilities;
- negotiation and consummation of arrangements with concessionaires, licensees, tenants and other intended users of the lobby, retail and other facilities within the Hotel;

- recruiting, engaging and training the initial staff of the Hotel;
- executing pre-opening festivities for the Hotel;
- preparing maintenance manuals for Hotel building systems and testing those systems;
- applying for and obtaining all licenses and permits necessary for the operation of the Hotel;
- providing pre-opening monthly reports and statements. The better practice is for the parties to agree on a detailed pre-opening budget providing, on a line-item basis, the identity of all salaries and wages, sales and marketing, and other general and administrative expenses that will be incurred through and including the Hotel's opening. The parties should remain informed of pre-opening booking activities in order to reduce potential increases in pre-opening expenses which may result from cancellation penalties in connection with any delay in the opening of the Hotel.

Residences Marketing and Licensing Agreement

This agreement may not be applicable to every hotel transaction because it permits Owner to sell residential real estate that is "Branded" by the selected Hotel Brand that will operate the adjacent Hotel. Licensing fees vary and are based on the gross revenue derived from the sale of the residences, inclusive of any special charge for the FF&E package, less sales commissions and reasonable closing expenses. Depending on the price point of the residences, the license fee may be calculated from revenues after a minimum per square foot return to Owner, and in so doing, more accurately reflect a percentage of the "lift" or "bump" in sales price resulting from the Brand. More often than not, the parties will exclude from the license fee the value of any incentives given to the residence purchaser such as pre-paid HOA fees.

It will be important to properly identify the "marks" which will be permitted to be displayed in sales and marketing materials and on and within the residence buildings. The license agreement should explicitly permit the project developer's use of the license in its resale of residences. However, the agreement would typically not extend the license to the resale of residences by parties to whom the licensee-developer has sold a residence. The impact of this is that any resale by an individual residential unit Owner does not earn the licensor another license fee.

An important aspect of the Residences Marketing and License Agreement is its impact on the timing of residence closings. Branded Managers typically insist that no residence sales may be closed, and no resident take occupancy, unless and until the Hotel is open for business. Owners, however, wish to close on the sale of residences as residence floors are completed from the bottom up (assuming all ingress, egress and fire/life safety systems are in place). Being able to sell residences early and sequentially over time may permit Owner to realize substantial dollar savings through early retirement of the construction loan. Early and sequential residence closings may also foster a more orderly occupancy plan for a building housing both the residences and the Hotel; ideally, most residents would have taken occupancy before the grand opening of the Hotel. A frequent compromise is to permit residence sales as soon as the public space components of the Hotel, such as the lobby, restaurant, and fitness center are open for business.

The term of the license agreement should comfortably accommodate sell-out of the residences. Once the residences are sold, and with the exception of a resale of a residence by the licensee developer (as opposed to resale by an individual unit Owner), the agreement terminates, the residences remain "Branded" by Manager's "marks," and the agreement then becomes limited to the continued Branding of the residences and the management services offered by Manager to the residences. Management services may include the following:

- identifying the "Brand" name at the residences (but not using the Brand name in any of the condominium documents or declarations which legally create the residences)
- the provision of certain Hotel services such as room service, laundry, valet and housekeeping services
- property management services such as front desk, loading and receiving and common area maintenance. In the event the residences do not want Manager to manage Owners' association, the agreement with Manager may be limited to a trademark license agreement affording Branding to the residences and permitting residence Owners to avail themselves of basic and à la carte services provided at the Hotel or by Hotel staff to the residences, such as room service and housekeeping. In the event the Hotel Management Agreement is terminated for any reason, the license agreement will also terminate, including use of the "marks" at the residences.

The Hotel Management Agreement examined

In this section, we will dissect the Hotel Management Agreement ("Hotel Management Agreement") and explore its component parts in great detail. Sample language will be used to help guide the discussion or to make a point, but this is by no means an endorsement of any particular language or negotiating strategy. Each Hotel Management Agreement negotiation will bring with it unique elements and concerns, and the worst thing a negotiator can do is to rely upon form language as the basis for establishing a position.

The companion website at www.routledge.com/cw/migdal includes many forms and examples, including a form of Management Agreement that should be examined in the context of this chapter. The form will provide you with a glimpse at some of the language used within a Hotel Management Agreement and allow you to explore how elements of the document work together.

The parties

The parties to the Hotel Management Agreement are the Hotel Owner ("Owner") and the Hotel Manager (the "Manager"). This simple concept may not be as simple as it seems in the marketplace of today with Real Estate Investment Trusts, private equity firms and a variety of ownership entities that will play the role of Owner. The underlying structure of the transaction might have changed between the effective date of the LOI and the opening of negotiations of the definitive agreements, making it essential to identify Owner as well as Manager. This is going to be very important as the balance of the Hotel Management Agreement is developed and attention must be paid to various representations, warranties and covenants, as well as the indemnification obligations of both Owner and Manager. Owner and Manager must be able to make the statements attributable to them in the Hotel Management Agreement and to stand behind them. An Owner or Manager that is not inherently creditworthy may need to provide a guarantor or other device to financially secure its respective obligations under the Hotel Management

Agreement. For example, either Owner or Manager may be a special purpose entity with no assets other than this one Hotel, in the case of Owner, or this one Management Agreement, in the case of Manager. In that case, each of Owner and Manager is likely to seek from the other a guarantor or some other vehicle through which to secure the financial obligations of the other party to the agreement. There will be circumstances when use of a regional operating entity as Manager, on the one hand, and use of a new single purpose company as Owner, on the other, will be perfectly logical and make business sense. But when either Owner or Manager has limited assets, steps will generally be taken to support the obligations of that type of Owner or Manager. The lesson to be learned from this is to examine the parties carefully with no assumptions as to bona fides or net worth.

The hotel asset

Manager is being engaged to provide professional management to the Hotel for a fee. This makes it incumbent upon both parties to be very particular in describing the Hotel and the various components. Owners would like to be precise and say something such as: "Owner desires to engage Manager to manage and operate the Hotel for the account of Owner as part of the System (which would then be defined in the agreement), and Manager desires to accept such engagement, in each case upon and subject to the terms and conditions set forth in this Agreement." This configuration is indicative of a Manager that implicitly recognizes that Manager works primarily for Owner, but, in return, Owner recognizes that the Hotel will, unless it is an independent hotel, be part of the Brand's larger System of Hotels.

It is also important to describe in sufficient detail as is appropriate under the circumstances, the components of the Hotel. For example, typical components would include:

- the building, including a reference to the minimum number of guest rooms and suites, restaurant(s), lounge(s), conference and meeting rooms, and other amenities;
- mechanical systems and built-in installations of the Building including, heating, ventilation, air conditioning, electrical and plumbing systems, elevators and escalators, and built-in laundry, refrigeration, and kitchen equipment;
- furniture, furnishings, wall coverings, floor coverings, window treatments, fixtures and Hotel equipment, and vehicles (the "FF&E");
- Fixed Asset Supplies;
- stock and inventories of paper supplies, cleaning materials and similar consumable items and food and beverage;
- pools, fitness center, spa, and parking garage.

Term

Structuring the Initial Term of the Agreement and any Renewal Terms, as well as the criteria, if any, for renewal, should all be resolved during the negotiation of the LOI. The Term of the Agreement will vary wildly depending upon a number of factors, the most notable of which is whether the Hotel is being operated as an independent Hotel or managed or Franchised as part of a Branded system of Hotels. Most of the Branded Hotel operating companies have spent decades selling off their owned Hotels as part of the transformation of the company into a management company. The value of the modern

Branded Hotel management company lies not in the value of any real estate (unless, of course, it still owns some real estate, and that does happen from time to time), but in the consistent flow of income to the company generated by long-term Management Agreements. This underlying valuation driver is what leads to Branded Management Agreements being so long and so difficult to terminate prior to the natural expiration of the stated Term.

The Initial Term is then followed by a series of Renewal Terms. There is a tension between which party, Owner or Manager, should have the right to exercise each renewal option and if that exercise is simply in the discretion of one of the parties or predicated on some economic or performance threshold so that the right to renew is something Manager has to earn. For example, it would not be unusual to see renewal rights expressed in this manner:

> [Owner] [Manager] shall have the option to renew this Agreement for a single addi-
> tional term of ____ (_) years (the "Renewal Term") by giving written notice of
> renewal to [Manager] [Owner] not less than _____ (_) days before the end of the
> Initial Term.

Although a single Renewal Term is stated above to offer a simple example, it is more common for there to be multiple Renewal Terms, particularly when there is a Branded Manager. In an agreement with an Initial Term and one or more Renewal Terms, care should be taken to then make certain that the definition of the "Term" includes both the Initial Term and any applicable Renewal Term(s).

Appointment of Manager

The Management Agreement is the vehicle through which Owner engages Manager as the exclusive Operator of the Hotel. In a deeper sense, this component of the Management Agreement is intended to express the agreement of the parties that Manager will exclusively supervise, direct and control the management and operation of the Hotel from the first date that the Hotel opens to the public for paying guests (the "Opening Date") and thereafter for the duration of the Term. Of course, Owner's intention is that Manager discharges that obligation subject to the terms, conditions, provisions and limitations of the Management Agreement, including, among a significant number of other elements, the approved annual Budget. Manager's acceptance of this appointment and agreement to manage the Hotel as stipulated by the Management Agreement should also be memorialized in this section of the Management Agreement.

Compliance with standards

A Management Agreement should express some minimum standard for the conduct and behavior of Manager. Both Owner and Manager will understand that the Hotel's compliance with a standard will be an important factor in the financial and operational success of the Hotel. Exactly what the standard will be and how it will be memorialized in the Management Agreement is subject to variations based on being part of a system of hotels under the umbrella of a Brand, rather than being an independent Hotel. If there are elements of the project that are not intended to be operated by Manager because those elements are subject to operation under a lease or with another manager, as might be the case with separate food and beverage venues, a fitness center, spa or retail stores, those operations must nevertheless comply with the standards.

Flexibility should be incorporated into the Management Agreement, particularly if it is a long-term agreement, to accommodate changed circumstances over time. For example, on the Management Agreement's effective date, Owner and Manager might intend to have all food and beverage areas associated with the Hotel operated by Manager. Despite these good and clear intentions, that may change and Owner may want to engage a third-party restaurant Operator for all or some portion of the food and beverage operations. The Management Agreement should specify how a decision like this is to occur. Is it Owner's decision alone, or will the parties be obligated to mutually determine that any or all of the food and beverage areas be operated by a reputable third-party restaurant operator under a lease, concession agreement, management agreement, franchise agreement, or similar agreement?

At certain higher room rate price points or within certain properties that are at the higher end of the spectrum of hospitality quality and service segments, Owner and Manager may be best served by mutual collaboration and agreement as to the business plan, strategy, conceptual development, and any other aspects regarding the implementation of any food and beverage operations within the Hotel. The lease, concession agreement, management agreement, franchise agreement, or similar agreement with respect to food and beverage operations will then be subject to a series of other requirements. Some of these requirements should include:

- consistency with the terms of the Management Agreement;
- consistency with the Standards;
- Manager's review of the document to confirm that it meets the requirements of the Management Agreement;
- all specific operating requirements such as providing in-room dining to Hotel guests or required menu items;
- all training requirements.

Because both the initial compliance with System Standards and the continued maintenance of System Standards can be costly, the Management Agreement can be negotiated to provide limitations on Owner's obligation to implement or fund revisions made to the System Standards within a short period of time after the Effective Date, particularly when the Hotel will be either new or newly renovated as of the Opening Date in compliance with the System Standards. The approach to this issue is not to limit Manager's right to modify its System Standards. Manager, particularly the Branded Manager, will have the unfettered ability to modify its System Standards as and when it deems it appropriate to keep the Brand fresh and to maintain a competitive edge. Instead, the approach is to limit or phase in Owner's obligation to immediately comply with changes in System Standards as applied to Owner's Hotel. Equally important will be the separation of potential changes in System Standards among various categories such as structural elements, life/safety requirements, requirements to maintain legal compliance, and other changes of a less crucial nature. It would not be unusual to have a range of time periods, and for Owner to have a "pass" on the immediate implementation of certain changes, while also providing a longer period for structural changes, a shorter period for technology upgrades and immediate compliance for life/safety and legal requirements. This attention to System Standards has benefits for both Owner and Manager. Manager, and particularly the Branded Manager, will impose certain System Standards upon Owner and the Hotel, with the ability to enforce those standards as well as modify them to continuously keep the brand up to date and fresh. Owner will know that all other Hotels within the Brand System are equally obligated to insure compliance with the System Standards, which can have long-term positive effects to every Owner. Owners will also seek to protect against any deterioration of the Brand Standards.

One strategy is to include within the Management Agreement an Owner's right to terminate in the event that Manager adopts or seeks to adopt System Standards that are lower than those initially or currently applicable to the Hotel. To Owners of Branded Hotel properties, the System Standards gives comfort that all hotels in the system will be required to meet the same standards and guests enjoying the experience in any one location will seek to duplicate that experience in other locations of the same Brand. If there is a lack of Brand consistency or deterioration in System Standards, an exit from the System should be available to Owner.

Management of the Hotel

Owner's appointment of Manager will be, in a general sense, to act pursuant to the Management Agreement, the System Standards and the approved Annual Budget. The Management Agreement will go deeper than that and specify certain management functions and services within the sole purview of Manager, as well as those functions and decisions that would require either consultation with Owner or the approval of Owner.

Some of the more customary functions, particularly in connection with a Branded Manager include those added below. An important note is that Manager's obligation to discharge all of these and any other management functions is predicated on the Hotel having sufficient Working Capital to permit Manager to discharge its duties or Owner providing the necessary Working Capital to allow Manager to address any shortfall of funds from Gross Revenue. The cost to discharge Manager's duties is an Operating Expense of the Hotel and would not be paid for by Manager from Manager's own funds. Customary functions might include:

- manage the Hotel in compliance with the operational aspects of the System Standards, to conduct the operation of the Hotel in a proper, orderly and business-like manner and to maintain the quality of the Hotel;
- recruit, employ, train, supervise, direct and discharge the Hotel Employees, and determine compensation;
- establish prices, rates, and charges for services provided in the Hotel, including room and suite rates;
- develop, revise and implement policies and practices relating to all aspects of the Hotel, which may be included in policy manuals;
- develop policies and practices to specifically address the purchasing of supplies and services (this might include competitive bidding procedures), the control of credit, and the scheduling of maintenance;
- manage expenditures to replenish Inventories and Fixed Asset Supplies;
- coordinate the replacement and procurement of FF&E;
- manage payments on accounts payable and collections of accounts receivable;

- prepare a Marketing Plan and Marketing Budget, and develop, arrange, and implement advertising, marketing, promotion, publicity and other similar public relations programs for the Hotel;
- prepare and deliver interim accountings, annual accountings, the Annual Budget, Building Estimates, Repairs and Equipment Estimates, and other Owner or Mortgagee requested information;
- plan for, execute, and supervise repairs and maintenance of the Hotel;
- negotiate licenses, leases, subleases, and concessions for the use of commercial space and meeting facilities at the Hotel;
- provide, in consultation and cooperation with Owner, risk management services relating to insurance;
- obtain and keep in full force and effect all licenses and permits (such as the very important liquor licenses);
- collect, account for, and remit to the governmental authorities all applicable excise, sales and use taxes or similar governmental charges collected by the Hotel directly from patrons or guests such as gross receipts, admission, entertainment, cabaret, use or occupancy taxes, or similar or equivalent taxes;
- meet with Owner's representatives regularly to review and discuss the previous and future Monthly Report, cash flow, potential major Capital Improvements, important personnel matters, marketing, strategic initiatives, events, variances from the approved Annual Budget and any other general concerns of Owner or Manager;
- reasonably cooperate with Owner in connection with a sale of the Hotel or to comply with any Mortgage;
- receive, consider, respond to and resolve all complaints and problems of any Hotel guest, customer, vendor or tenant in a timely and effective manner;
- comply with any covenants, conditions, and restrictions recorded against the Hotel;
- use commercially reasonable efforts to operate the Hotel in accordance with any Mortgage;
- enter into contracts or purchase agreements for purchases or services related to the Hotel, subject to a monetary limitation.

Compare this list of management functions and duties with those in the following list where Manager may have no or limited authority to act without the prior consent of Owner. Customary functions that require the prior consent of Owner might include:

- entering into a lease, license, contract, or other arrangement (or series of related contracts or arrangements), even if it is in the approved Annual Budget, if the expenditure exceeds an agreed upon cost limitation, if the term of the lease, license, contract, or other arrangement exceeds an agreed upon period of time or if the relationship with the vendor cannot be terminated by Owner without payment of a termination fee;
- settling certain insurance claims or condemnation claims, which also may be subject to a cost limitation;
- instituting, defending or settling legal or equitable proceedings with respect to the Hotel, including the selection of counsel, but possibly subject to limitations as to cost or the nature of the claim;
- entering into a transaction with an Affiliate of Manager;
- acquiring, on behalf of Owner, any land or interests in land;

- financing, refinancing, mortgaging, placing any liens on or otherwise encumbering the Hotel, the Building, or the Site;
- selling all or any portion of the Hotel except for dispositions of food, beverages and merchandise, and dispositions of FF&E, in the ordinary course of Hotel business and consistent with industry standards;
- entering into any collective bargaining agreements with respect to any Hotel employees, or recognizing a collective bargaining representative to act on behalf of any Hotel Employees;
- making any press releases or public announcements regarding the Hotel or Owner, other than any agreed to press releases or announcements for marketing and promotion of the Hotel;
- undertaking transactions involving expenditures from the FF&E Reserve greater than as agreed on in the approved annual operating budget;
- undertaking transactions involving capital expenditures greater than an agreed on threshold, even if undertaken to comply with a new or amended Brand Standard.

Anytime an Owner finances a portion of the cost to acquire the Hotel and subjects the Hotel to the encumbrance of a Mortgage, Manager's obligations will include compliance with the Mortgage, so long as Owner has provided Manager with a copy of the Mortgage or, and as preferred by Managers, complete and accurate summaries of the relevant provisions of the Mortgage, and the provisions of the Mortgage and compliance with those provisions by Manager are applicable to the day-to-day operation, maintenance and non-capital repair and replacement of the Hotel, do not require payments of Manager's own funds, do not materially increase Manager's obligations under the Management Agreement or materially decrease Manager's rights under the Management Agreement, do not limit or purport to limit any corporate activity or transaction with respect to Manager or its Affiliates or any other activity, transfer, transaction, property, or other matter involving Manager or its Affiliates other than at the site of the Hotel, and are otherwise within the scope of Manager's duties under the Management Agreement.

Because Manager's ability to comply with the terms of a Mortgage will be heavily dependent on Owner, the Management Agreement will include an Owner acknowledgment or covenant that it would not be deemed a breach by Manager of its obligations under the Management Agreement if Manager or the Hotel fails to comply with the provisions of any Mortgage because of circumstances and events such as:

- the condition of the Hotel or the failure of the Hotel to comply with the provisions of a Mortgage or preexisting recorded agreement or covenant prior to Manager assuming the day-to-day management of the Hotel;
- Owner's failure to provide adequate required Working Capital;
- construction activities at the Hotel;
- inherent limitations in the design and/or construction of or location of the Hotel;
- written instructions from Owner telling Manager to operate the Hotel in a certain manner.

Under all circumstances, Owner remains ultimately responsible for the burdens of ownership of the Hotel, and Manager is not obligated to spend any of its own money to discharge ownership obligations. As it relates to compliance with Owner's mortgages and dealing with Mortgagee, Manager may agree to cause the Hotel to remain in compliance with any Mortgagee mandated obligations as to the physical plant and operations, but only so long as the cost of that compliance is paid for by Owner.

Owner and Manager have obligations to ensure that the Hotel and its operations comply with Legal Requirements that may go beyond any Mortgagee requirements. These requirements will capture everything from the design and construction of the physical plant of the building and its major operating systems, which are primarily Owner's responsibility as related to ownership of the building, to the daily functioning of the Hotel, which are primarily Manager's responsibility as related to operations rather than ownership. If, despite Manager's exercise of commercially reasonable efforts to comply with Legal Requirements, Manager receives notice that the Hotel does not comply with Legal Requirements, Manager must act promptly to take those steps that are reasonably necessary, in concert with the Hotel's legal counsel and Owner, to remedy the non-compliance. While Manager may not have a legal obligation to do so, Manager will have the right to take appropriate steps, at Owner's expense, to comply with, or cure or prevent the violation of, any Legal Requirements, avoid or minimize any actual or potential injury to persons or damage to the Hotel or other property, and avoid or minimize any risk of criminal or civil liability of Manager and its Affiliates.

Whether examining Mortgagee obligations or Legal Requirements, Manager's duties and obligations to Owner commence as of the effective date of the Management Agreement, and Manager has no liability or responsibility whatsoever for any conditions existing as of the assumption of management by Manager. The Management Agreement often includes representations and warranties in favor of Manager from Owner confirming that as of the effective date of the Management Agreement, no conditions or claims exist with respect to the Hotel. If Manager incurs any cost or expense in connection with a prior condition of claim, it is most often treated as either Operating Expenses or Capital Improvement of the Hotel, but then excluded from Operating Expenses and not treated as Capital Improvements for the purpose of calculating or determining the Incentive Management Fee, the Operating Profit Test, and the RevPAR Test. The end result of all of this is that Manager is not negatively impacted as a result of a claim that may have existed or a condition at the Hotel that existed prior to the effective date of the Management Agreement.

But what if there was a prior condition that first comes to light during the Term? As an initial matter, any work relating to that prior condition is beyond the scope of the Management Agreement. Manager's services under the Management Agreement do not extend to management of any remediation, abatement or other correction of those prior conditions unless specifically raised by Owner and accepted by Manager. That leaves Owner with complete responsibility and liability for the matter. Owner will have to discharge that responsibility, but because this is an operating Hotel under the exclusive management of Manager, Owner's actions must be discharged with as little disturbance or interruption of the use and enjoyment of the Hotel as practicable. Although Owner and Manager will have some alignment of interests in the prompt correction of the prior condition, Manager will be very interested in preserving the integrity and proper performance levels of the delivery of all guest facing services and amenities to the Hotel's guests.

System Services

System Services is a term that is intended to describe the variety of services that Manager either performs or arranges for others to perform at the Hotel for which no fee is charged beyond the Base Management Fee. When an Owner asks a Manager, "What does my Base Management Fee pay for?", the response is System Services.

An illustrative list of System Services follows, but it is by no means an exclusive list and it is subject to good faith negotiations between Owner and Manager in the development of the Hotel Management Agreement:

- system financial planning and policy services;
- product planning and development;
- human resources management and planning for the System;
- development and implementation of Manager's technical and operational programs;
- corporate planning and policy services;
- financial planning and corporate financial services;
- corporate executive management;
- protection of Manager's Trademarks.

Centralized Services

Centralized Services is a term that is intended to describe those services that Manager will perform, either directly or through an Affiliate, for the benefit of the Hotel. Centralized Services are included within the approved Annual Budget to provide Owner the opportunity, on an annual basis, to examine and evaluate the schedule of services that Manager intends to deliver on a centralized basis, and the cost of the services to be provided. Manager or Manager's Affiliates that provide Centralized Services will be reimbursed for the Hotel's fairly and equitably allocated share of the total costs that are actually and reasonably incurred in providing the Centralized Services on a system-wide basis to Hotels managed by Manager or its Affiliates. It is the inherent nature of Centralized Services that they are more efficiently delivered on the basis of the entire system of Hotels that benefit from the services rather than delivering the services to any single Hotel. The cost of delivering Centralized Services, for which Manager is entitled to reimbursement, is an allocation of aggregate costs. These costs may include salaries (including payroll taxes and employee benefits) of Corporate Personnel directly providing the Centralized Services and the cost of all equipment employed exclusively in providing the Centralized Services to the Hotel and not also to other Hotels in the System.

The range of services typically included within Centralized Services includes the following:

- accounting services;
- purchasing services;
- group benefits and services;
- payroll and benefits services;
- financial services;
- revenue management services;
- engineering services;
- information technology;
- business intelligence;
- risk management;
- field marketing;
- sales and public relations;

- human resources;
- on–site sales training;
- associate satisfaction surveys;
- Manager's training program and other training.

The cost to the Hotel for receiving Manager's Centralized Services is determined by calculating the Hotel's share of the costs as a Hotel within the System. This may be modified on occasion to address only a part of the System, such as segregating resort properties from the balance of the System. Those costs should be determined in a fair and equitable manner by Manager. Within a large system of Branded Hotels, there will need to be a degree of trust between Owner and Manager, and a meaningful exchange and dialogue during the process of annually developing the approved Annual Budget for the Hotel, because it is often difficult for Owner of any single Hotel or group of Hotels within the System to look into Manager's precise methodology for determining the costs of Centralized Services. The cost of Centralized Services is an Operating Expense of the Hotel and borne by Owner and paid or reimbursed to Manager out of the appropriate Operating Account of the Hotel or, if the amount on deposit in the Operating Account is inadequate, by Owner. The Management Agreement will include a system for Manager to make demands on Owner for funds, and a period of time during which Owner must reimburse Manager or risk being in default under the Management Agreement.

Because so much of the content and delivery of Centralized Services is subject to Manager's discretion and control, many Management Agreements include representations and warranties from Manager for the benefit of Owner about Manager's program. In theory, should Manager be using the Centralized Services program as a profit center or fail to allocate the costs in a fair, equitable and reasonable manner, Owner has a foundation in the Management Agreement to seek redress from Manager. Manager will often represent and warrant to Owner that:

- Manager has disclosed to Owner the types of Centralized Services that Manager currently makes available to System Hotels and the costs that would be allocated to the Hotel;
- the Hotel is expected to receive substantial benefit from its participation in Manager's system for the delivery of the Centralized Services.

Owner, for its part, acknowledges and agrees that:

- Manager is entitled to payment for its delivery of the Centralized Services in addition to its Base Management Fee and Incentive Management Fee;
- Manager's receipt of payment for its delivery of the Centralized Services, in and of itself, does not breach any fiduciary or other duty that Manager may have to Owner.

Owner can usually negotiate for the right, in its discretion, to annually audit the delivery of Centralized Services and the aggregate and allocated costs charged by Manager.

Marketing/sales contribution

Branded Hotel Managers will have a system of Brand Marketing Services and a national sales program as part of their System. Like Centralized Services, this National Sales Program is to be operated in a fair and equitable manner to benefit all Hotels in the System, and no single Hotel or group of Hotels, including Hotels owned by Manager or its Affiliates, should receive a disproportionate benefit from or be allocated a disproportionate cost of the Brand Marketing Services and National Sales Program.

Owner is obligated to pay a fair and equitable share of the costs for the Brand Marketing Services and the National Sales Program, and many Owners specify in the Management Agreement a maximum amount or "cap" on the annual cost.

Reservation Systems and Fees

One of the primary benefits to an Owner of engaging a Branded Hotel Manager and affiliating with a Brand instead of being an independent Hotel comes from the benefits Owner receives by being part of a Reservation System. This is a benefit that some Brands may offer without charge, but that typically comes at a separate cost to Owner, most often, through a per reservation fee. The Reservation Fee of Manager's Reservation System is subject to change, and the charge is expressed as a dollar amount per reservation booked through Manager's corporate reservation system. Manager will change its fee as appropriate annually, but the change must apply uniformly and in a non-discriminatory manner to all Hotels in the System.

The delivery of services through a Reservation System may be memorialized and expressed within the Management Agreement, but it need not be that way. Some of the larger and more developed and robust Branded Hotel Managers and Management Companies provide reservation services and a Reservation System through a separate, but affiliated company and pursuant to a separate Reservation Services (or similarly named) Agreement.

A good general understanding of what "Reservation Services" is intended to include contemplates all manner of electronic and voice reservation services provided by Manager and/or its Affiliates through telephone reservations arranged through call centers, regardless of where the call center is located, reservations through the internet, or through any number of Global Distribution Systems such as Amadeus/System One, Apollo/Galileo, Sabre (Abacus), and Worldspan, which may involve additional charges for these systems.

The Reservation Services provided to Owner include:

- the maintenance of the computers and equipment;
- staffing the call centers;
- development activities to support and improve the Reservation Services.

Reservation Services is typically not a profit center for Manager. Instead, the Hotel is required to reimburse Manager or its Affiliate providing the Reservation Services for the cost of providing Reservation Services based on a sharing formula. The formula will vary from provider to provider, but a general formula would be based on the Hotel's pro rata share of all costs incurred by Manager (or its Affiliate) in providing the Reservation Services to all Hotels within the System or to All Hotels using the Reservation Services.

What would be subject to negotiation in this regard would be the basis of the allocation. For example, would the allocation be based on a per room basis, number of owned Hotels in the Brand's system or something else?

It is common to see this expressed as the average costs incurred by Manager in providing the Reservation Services on a per reservation basis. The agreement then goes on to express the cost per reservation, plus a charge for each reservation that is booked through Manager's proprietary Reservation System, as well as local reservations made at the Hotel or a shared service center which are transacted through Manager's proprietary Reservation System.

In this situation, the charge per reservation transaction is assessed without regard to the number of rooms or the number of nights booked in each reservation, and Owner acknowledges that the cost and reimbursement formula is subject to change. Owner just needs to be sure that all pricing adjustments are communicated to Owner prior to going into effect, usually at least 30 days prior to the effective date of the change, and that all Hotels within the Reservations System are subject to the same change.

If there is a separate Reservation Services Agreement, it will contain many of the standard provisions of the Management Agreement itself to render the Reservation Services Agreement a fully integrated, separate and freestanding agreement with Owner, with a term that is co-terminus with the Management Agreement. Both Manager and Owner will want to confirm that these provisions in the Reservation Services Agreement conform to the same or similar provisions in the Management Agreement.

Although the Hotel will be obligated to use Manager's Reservation System as its sole vehicle for reservations, Manager (or its Affiliate) will make certain covenants to Owner and have certain obligations to Owner. These would typically include the following:

- to provide current information about availability and rate for the Hotel;
- to include the Hotel in the Reservation System;
- to identify all transactions transmitted to the Hotel through use of an identification code that will identify the origination of the transaction;
- to supply Owner with a report showing bookings, cancellations, and other additional information that Owner and Manager may have negotiated in the Reservation Services Agreement, usually prior to the last date of each month for the immediately preceding month;
- to provide other information to Owner that is reasonable under the circumstances, substantiating the Reservation Services charges;
- to provide to Owner the information Owner will need to obtain any required hardware and software for the Reservation System, all of which is an Owner expense;
- to use commercially reasonable efforts to minimize interferences with, or diminishments or interruptions in, the Reservations System;
- to not knowingly take action to preclude or in any way impair, the ability of users to book reservations at the Hotel though the Reservation System;
- to not divert, or attempt to divert, potential guests to other hotels;
- to diligently discharge its obligations without preference to any other Hotel;
- to use commercially reasonable efforts to maintain Manager's equipment and systems so as to perform its obligations with respect to Reservation Services, such as maintenance of back-up systems to minimize the risk of interruption.

Significant components and aspects of the delivery of Reservation Services by a Manager are not subject to Manager's control, such as internet or telephone service and connectivity. These types of agreements will include provisions to protect Manager from liability for these situations. Owner will covenant and agree that Manager will not be liable for any indirect, incidental, or consequential damages, including but not limited to, loss of revenues, to the extent arising other than from the gross negligence and/or willful misconduct of Manager. In addition, unless it is specifically negotiated by the parties and expressly included in the Reservation Services Agreement, Manager will make no warranty, express or implied, including any implied warranty of merchantability or fitness for a particular purpose, with respect to the Reservation System, the intellectual property or the software operating the system.

Purchasing services

An Owner can negotiate to utilize its designated purchasing agent in the Management Agreement, Pre-Opening Agreement and/or Technical Services Agreement in many situations where there is not a national Branded operating company in place, and even with a national Brand in certain circumstances. If Owner does not have the opportunity to designate the purchasing agent or does not take advantage of that opportunity, Manager will make available to Owner the services of Manager's corporate purchasing personnel and facilities for purchases of Fixed Asset Supplies, FF&E and other goods and services required for the operation of the Hotel under the general category of "Purchasing Services."

Owner has a decision to make. It will, when it has the opportunity, have to either bear the sole responsibility for ensuring that all goods and services that it purchases for the Hotel, other than those selected or approved by Manager, comply in all material respects with System Standards, or use the Purchasing Services of Manager.

When Manager is one of the larger Brands, it is not unusual to see Avendra identified as the provider of purchasing services and various elements of supply chain management. The Hotel Owner is generally not going to assume the role of purchasing agent, so if the provider of those services provides the services professionally and at market competitive prices, Owner can have meaningful choices in whom to engage.

If Owner elects to use Manager's Purchasing Services, Manager is entitled to compensation that Manager discloses to Owner. In addition, when taken as a whole, including consideration inuring to the benefit of Manager and its Affiliates, the cost of Manager's Purchasing Services must be as favorable to the Hotel as the prevailing terms of contracts to provide similar goods or services on a single-property basis obtainable on a commercially reasonable basis from unrelated parties in the geographical area of the Hotel. The Management Agreement will typically be very clear whether this will be tested on a line-by-line basis or in the aggregate. Manager's Purchasing Services cannot be out of line with what Owner could achieve for its single Hotel in the marketplace. This helps align Owner's and Manager's interests and priorities and assures Owner that the Hotel will not overpay for what it purchases from Manager.

Manager's compensation for Purchasing Services when part of Manager's program for the System is on terms that are applied uniformly and in a non-discriminatory manner to all Hotels in the System and generally disclosed in advance to and approved by Owner. Owner may also attempt to negotiate a fee "cap" based on a percentage of the third-party costs of the goods and services of the purchases.

Another possible variation is to segregate purchases from "Major Purchases." More routine and mundane purchases might be the purview of Owner, while Major Purchases would be managed by Manager under Purchasing Services. The concept of "Major Purchases" attempts to capture those purchases that arise in connection with a substantial renovation of the Hotel, and therefore would be approved by Owner and provided for in an approved Capital Budget.

In addition to compensation in the form of a Purchasing Services fee, and subject to the approved Annual Budget, Managers are also entitled to reimbursement of Manager's actually incurred, reasonable administrative costs, without any mark-up or profit component, directly relating to the Purchasing Services for all other purchases for the Hotel made utilizing the Purchasing Services and provided for in the approved Annual Budget which do not constitute Major Purchases.

Some of Manager's purchasing functions involve competitive bidding. In those situations, when taking bids or issuing purchase orders, Manager is expected, and contractually obligated under the Management Agreement, to use commercially reasonable efforts to secure for, and credit to, Owner any Rebates obtainable as a result of the purchase. The Management Agreement will provide that in any instance in which Manager receives a Rebate with respect to any purchase, sale, lease or other procurement or provision of goods, services, systems or programs for or to the Hotel, Manager must promptly pay the Rebate to the Hotel. In the event that any purchase is for the benefit of the Hotel and other Hotels operated by Manager or its Affiliates, Manager must pay to the Hotel a share of the Rebate. That Rebate share is calculated on a fair, reasonable, and non-discriminatory basis. This system of checks and balances is intended to ensure that neither Manager nor any Affiliate of Manager receives, directly or indirectly, any remuneration other than that to be paid by Owner to Manager under the Management Agreement. Many Brand Managers will make the point that any Rebate should be applied first to defer or offset the cost of the operational costs of the purchasing program. This is neither illogical nor unreasonable, so long as the record keeping is transparent and provides Owner with the necessary data on purchasing to track the allocation of the Rebate.

Additional services

Management Agreements are well devised and negotiated documents, but due to their long-term nature should not be expected to address every service that the Hotel might ever require or that a Manager might be able to offer to an Owner. For this reason, provision is made to permit Manager to make available to Owner from time to time any other services that Manager can provide for the benefit of the Hotel. These services and the consultants that may provide them if not directly provided by Manager will be provided for in the Annual Budget. If an Owner wants Manager to provide additional services, Owner and Manager negotiate in good faith the terms and conditions under which the services shall be provided.

Owner's right to inspect

Despite the degree of trust between Owner and Manager that can be very helpful in a Management Agreement context, Owner should retain the right to have access to the Hotel for the purpose of inspection of the Hotel, Hotel operations and Hotel books and records or, upon reasonable advance notice to Manager, for the purpose of showing the Hotel to prospective purchasers, tenants, investors or Mortgagees. Owner, in its interest and well as in the interest of Manager, should endeavor to minimize disruption to the Hotel and its guests when Owner is at the Hotel. Another way Owners keep watch over what is happening at the Hotel is through the engagement of a professional hotel asset manager. The asset manager works for Owner and is generally paid by Owner. Hotel Owners that own Hotels as a class of real estate among a portfolio of real estate holdings that also include office buildings, apartment buildings and shopping centers use asset managers to fill the gaps in Owner's knowledge base about Hotels and hotel operations, and to make sure that the Hotel and Hotel Manager are monitored when Owner is unable to do so.

Relationship of Owner and Manager

Management Agreements vary as to the manner of describing the relationship of Owner and Manager. Some Management Agreements use independent contractor language, while others use agency language, and some remain silent on the issue.

Recent litigation has clarified, at least in New York and Florida, that Hotel Management Agreements are agreements for personal services, but the majority of Management Agreements continue to use the language of agency. Chapter 4 will provide an analysis of some of the leading Management Agreement Cases.

Under agency concepts, in the performance of its duties as Manager of the Hotel, Manager acts as agent of Owner. This casts Manager as acting on Owner's behalf and for that reason, all debts and liabilities to third persons incurred by Manager in the course of its operation and management of the Hotel in accordance with the provisions of the Management Agreement are the debts and liabilities of Owner only, and Manager is not liable for any obligations by reason of its management, supervision, direction and operation of the Hotel in accordance with the provisions of the Management Agreement as agent for Owner. Manager is permitted to inform third parties with whom it deals on behalf of Owner that Manager is merely Owner's agent and not a principal.

Employees

There is some variation between domestic transactions for Hotels in the United States and Hotels located outside the United States. Although recent trends indicate some variation from past practice, in the case of Branded U.S.-based Hotels, it is common for all personnel employed at the Hotel to be employees of Manager (or one of its Affiliates) and not of Owner or its Affiliates. Variations on the identity and structure of the employer will arise in response to collective bargaining agreements that may be applicable to the Hotel and the overall absence or presence of one of more collective bargaining units.

Regardless of which party is the technical employer of the employees, Manager has the discretion with respect to the Hotel Employees, including decisions regarding hiring, promoting, transferring, compensating, supervising, terminating, laying off, assigning, controlling, directing, and training all Hotel Employees and establishing and maintaining all policies relating to employment of Hotel Employees. This is a significant amount of power and discretion in an area that is often one of the largest components of the Annual Budget and that, on a human level, can make or break a Hotel if it succeeds or fails to consistently deliver excellent services to its guests. This is why the Management Agreement will go into greater depth about how Manager will discharge this duty.

Manager's selection of Hotel Employees must be in accordance with System Standards when Manager is a Branded Manager or a Manager that is associated with a System.

In addition, Owner will want the right to interview and approve persons holding the most senior and strategically important positions at the Hotel, such as the Hotel general Manager, director of finance/comptroller, director of sales and marketing, director of revenue management, and director of food and beverage. Some Owners will also want the right to participate in the decision to terminate any senior executive personnel at the Hotel. From Manager's perspective, this can be problematic because an Owner might want to terminate someone for a variety of reasons both related and unrelated to hotel operations. Hotel Managers invest a significant amount of time and money to the training and development of senior employees, and their termination would have to be consistent with overall corporate policy and not just Owner's desires, unless clearly supportable through objective criteria and support.

Owners and Managers generally share the desire to populate the Hotel with only the most competent staff, but there is certainly room for disagreement when the question is termination. Managers will need to exert significant control in this area, but if the parties are prepared to add an Owner's right to participate in termination decisions to the Management Agreement, a procedure for dispute resolution should also be added. When the Hotel Manager is not a Brand and the Hotel is not subject to a collective bargaining agreement, Owner can have greater influence as to termination. Persons selected for these positions are generally subject to approval by Owner, although Owner must be reasonable and cannot withhold approval unreasonably. One common procedure for the hiring of the Executive Personnel or key employees is that Manager will notify Owner in writing of each candidate selected by Manager for an Executive Personnel position and will provide Owner with the candidate's résumé and credentials. Owner would then have the right to interview each candidate selected by Manager for an Executive Personnel position at a mutually acceptable time and place within a set period of time, often, 10 days following delivery to Owner of the candidate's resume and credentials. Owner must provide written approval or disapproval of each candidate, including a written explanation and basis for any disapproval, within a short period of time, often, five days following Owner's interview of the candidate or Owner's election to not interview the candidate, as the case may be. In the event that Owner does not provide a written approval or disapproval of any candidate within 15 days following delivery to Owner of the candidate's résumé and credentials, the candidate is deemed approved by Owner. In the event that Owner disapproves of a candidate, Manager will then submit one or more additional candidates for Owner's approval until the parties mutually agree on a candidate. During the vacancy of any Executive Personnel position, Manager has the power and authority to utilize Corporate Personnel to temporarily fill the vacant position, in which case the cost of that Corporate Personnel is an Operating Expense.

Some Management Agreements will limit Owner's approval or disapproval rights to three candidates, and if there is no approval at that time Manager may have the right to appoint a candidate to the vacant position. Having that candidate come from one of the three previously disapproved by Owner or specifically not one of the three previously disapproved by Owner is a matter of negotiation between the parties. The essence of negotiating this element of Owner–Manager relationship often comes down to diverging priorities. Manager, be it a Brand or not, benefits by offering employees a career path and the opportunity for advancement within Manager's System and the hotel industry. Advancement often requires relocation, and Manager will want the ability to transfer or relocate employees within the Brand System. Owner benefits from having well trained and experienced senior employees and reacts strongly to the possibility that

Owner's Hotel will be a training facility for Manager, where employees can gain valuable knowledge and experience only to move on to another Hotel. The departure of Executive Personnel negatively impacts the Hotel on both an economic level and on an overall morale and continuity level. Some Owners curb Manager's desire to relocate employees by shifting the cost to Manager, and as neither an Owner expense nor an Operating Expense, if certain employees are relocated at the initiation of Manager rather than the employee during a negotiated period of time after first being added to the Hotel payroll.

Manager is responsible for developing and implementing prudent policies and practices relating to Hotel Employees and their employment by Manager at the Hotel. Some typical components of these types of policies include:

- terms and conditions of employment;
- applicant screening and background checks;
- employee selection, hiring, training, and supervision;
- compensation, bonuses, severance, pension plans, and other employee benefits;
- employee discipline, dismissal, transfer, and replacement;
- the exercise by any Hotel Employees of rights under the National Labor Relations Act, the WARN Act, or any applicable labor laws including union organizational efforts, recognition or withdrawal of recognition;
- representation elections;
- contract negotiations;
- the determination of an appropriate bargaining unit or units;
- whether to negotiate on a single-employer, coordinated or multi-employer basis;
- grievances, unfair labor practice charges, strikes, boycotts or other economic activity and lockouts;
- compliance by Hotel Employees with Legal Requirements, including anti-discrimination, sexual harassment, safety and health, hazardous materials, and other environmental laws and any applicable collective bargaining agreements.

Under all circumstances, Manager is expected to, and is obligated under the Management Agreement, use best efforts to comply with all Legal Requirements in all matters with respect to the Hotel Employees. This will include matters such as decisions regarding hiring, promoting, transferring, compensating, supervising, terminating, directing, and training all Hotel Employees. In all of these matters, Manager's failure to comply will constitute an event for which Manager is obligated to indemnify Owner under the indemnification provisions of the Management Agreement.

Because the Executive Personnel are so crucial to the financial success of the Hotel, the Management Agreement will usually prohibit Manager from entering into any contract or other agreement for the employment of the Executive Personnel without the prior consent of Owner. Stronger Owners with enhanced bargaining positions may seek input, if not approval, for the hiring of all Hotel Employees. This is generally not achievable, particularly with Branded Hotel Managers, but it is a consideration of the Management Agreement. In addition, there is also significant negotiation concerning how many Executive Personnel will be subject to Owner's approval. In many situations with strong national Branded Managers, Owner's approval is limited to the General Manager. From that opening position, Manager and Owner should discuss how much deeper Owner's approval should extend. For Executive Personnel, it will probably extend after the termination or expiration of the Management Agreement to prohibit Owner from soliciting Manager's Executive Personnel for a period of time after the termination or expiration date of the Management Agreement.

Managers also seek discretion and latitude with respect to other benefits for their employees. Managers negotiate for the ability to provide lodging for Corporate Personnel visiting the Hotel in connection with the performance of Manager's services as well as for employee rates as a benefit of employment with that Manager. These arrangements allow employees of Manager to use Hotel facilities, subject to availability and without displacing paying guests. In addition to these types of "perks," there is the more serious matter of when the general Manager or other Executive Personnel may need to live at the Hotel. This is particularly important in new openings or the relocation of a new general Manager to the Hotel. Manager will negotiate for the Management Agreement to provide the general Manager of the Hotel and the other Executive Personnel temporary living quarters within the Hotel and the use of all Hotel facilities, in either case at a discounted price or even without charge. This arrangement will be time limited, such as 90 days for the general Manager and 30 days for all others. Owners often negotiate for similar benefits for Owner's agents, employees, friends, and family. Owner may have the right to reasonably direct that Manager provide complementary accommodations and services to Owner's guests and invitees, subject to availability, and without displacing paying guests. Other than rights and privileges specifically set forth in the Management Agreement, Manager will be prohibited from providing any other person with complementary accommodations or services without the prior approval of Owner. This is often expressed as a blanket prohibition as to Manager, except as may be provided for in the approved Annual Budget, for legitimate business purposes relating to the Hotel and otherwise in accordance with usual practices of the Hotel and travel industry. In each Monthly Report Manager delivers to Owner, Manager will include a reasonably detailed description of all complementary accommodations and services provided by Manager at the Hotel for the previous Accounting Period.

All compensation (including wages, fringe benefits, and severance payments) of the Hotel Employees, to the extent in accordance with the Management Agreement and the approved Annual Budgets will be an Operating Expense and therefore are borne by Owner and paid or reimbursed to Manager out of the Hotel's Operating Account or, if the amounts in that account are insufficient, by Owner within a very short period of time, such as five business days, following Manager's written demand for reimbursement. As a further limitation on Owner's authority and discretion, Owner does not have the right to disapprove Hotel Employees' compensation reasonably required to operate the Hotel consistent in all material respects with the System Standards. Manager has the right to institute severance payment policies and bonus programs for the Hotel Employees, but those policies and programs must be reasonable, competitive in the local market of the Hotel, customary in the industry, and consistent with the policies and programs for all other Hotels in the System and the terms of the Management Agreement. This is another area of Hotel management where compliance with System Standards will have a meaningful impact on Owner–Manager relationship.

To limit Owner's ability to hire Manager's employees, the Management Agreement will include Owner's agreement to not solicit the employment of the Executive Personnel at any time during the Term or within some negotiated period of time after the expiration or termination of the Management Agreement, other than continued employment at the Hotel.

There may be occasions during the Term when an Executive Personnel position becomes vacant and action must be taken in the near term while Owner and Manager are acting to fill the vacancy on a more permanent basis with a qualified applicant. If Manager elects to fill the gap temporarily using its Corporate Personnel to perform services at and for the direct benefit of the Hotel, then the cost associated with the use of Corporate Personnel should be promptly disclosed to and approved by Owner. If there was an opportunity for advanced planning or simply good thinking to anticipate this type of eventuality, those costs might be in an approved Annual Budget, but if that is not the case, Owner should approve the costs. The cost of temporarily placing Corporate Personnel at the Hotel will include the reasonable travel, lodging, meal, and other out-of-pocket expenses for the individual while working at and for the specific benefit of the Hotel. The purpose of this temporary use of Corporate Personnel is to serve the Hotel instead of a regular full-time equivalent Hotel Employee. It is not to have the individual housed at the Hotel serve other Hotels in the system, unless that is specifically disclosed and approved by Owner. It is often wise to limit the daily per diem rate for Corporate Personnel to the actual costs of Manager in providing its personnel at the Hotel, without mark-up or profit, although the cost could include the prorated portion of compensation and employee benefit costs. Another important point to address is that Manager should use commercially reasonable efforts to limit the cost of the reimbursable compensation and employee benefits attributable to Manager's Corporate Personnel to an amount not greater than the cost of the compensation and employee benefits that would otherwise be paid if a Hotel Employee were to fill the vacant position.

The requisite record keeping and paperwork for Hotel Employees generally falls to Manager as the employer or, if Owner is the employer, as part of the management services provided by Manager. Manager is required to maintain all personnel records and payroll systems, and to comply with all record keeping and reporting requirements of all state and federal laws relating to employees.

The termination of Executive Personnel is far more problematic than the Engagement of Executive Personnel. Although Owner may have certain rights to interview and approve applicants for Executive Personnel positions, those individuals remain, more often than not, employees of Manager. Managers are usually unwilling to give Owner any power to cause the termination of any Hotel Employee, including Executive Personnel. One often seen compromise position permits Owner to request, and the request must be reasonable, that the employment of a member of the Executive Personnel be terminated by Manager. The request must be made with good reason, making it reasonable, and supported by objective evidence of unsatisfactory performance or behavior by the individual. Manager is not obligated under the Management Agreement to terminate the individual, but must reasonably and in good faith consider Owner's request. If Manager is willing to act on Owner's request, Manager can terminate the employment of the individual or transfer the individual to another Hotel managed by Manager or any of its Affiliates.

Hotels can rise and fall on the strength of the employees who execute the business plan every day. This is one of the reasons that Owners try to minimize Manager's relocation of employees within Manager's system. Of course, individual employees often have a personal or professional need to relocate. The issue is actually a Manager relocating employees too frequently and not at the request of either Owner or employee. Relocation can disrupt not only the operations of the Hotel on the mundane, day-in and day-out level, but can also create a financial burden on the operating budget of the Hotel. The remedy is to have Manager assume some of the financial burden of certain employee relocations.

The Management Agreement can provide that if a Hotel Employee is relocated to another Hotel or property managed by Manager or any of its Affiliates within a negotiated period of time, such as 18 to 24 months after the employee's arrival at the Hotel and first being on the Hotel payroll, Manager will pay or reimburse Owner for part of the relocation expenses incurred in relocating that individual's replacement to the Hotel. It is not the cost of relocating the departing employee that is being addressed, because more often than not, that cost is being assumed by the Hotel to which that employee is going. Rather, it is the cost of relocating the new incoming employee to the Hotel that Owner wants to address. The amount to be paid to or reimbursed to Owner should equal the actual total relocation costs incurred by the Hotel in bringing the new Hotel Employee to the Hotel. Sometimes this is further refined by having the actual total relocation cost calculated proportionately to the required period of time. In the interest of fairness, Manager's reimbursement obligation as described here would not apply with respect to the relocation of any Hotel Employee whose transfer from the Hotel to another Hotel managed by Manager or any of its Affiliates was requested by Owner or by the Hotel Employee. Sometimes the Management Agreement can go one step further and make it clear that Owner will not incur any expense in connection with the relocation of any Hotel Employees to his or her new position at a Hotel managed by Manager or any of its Affiliates regardless of the length of time the individual is employed at the Hotel. This permits greater latitude to large national Branded Managers that have their own career paths and programs for their employees. If the cost is absorbed by Manager, Owner still has to deal with any disruption at the Hotel, but at least there is no cost to Owner and Manager should be attending to and compensating for any potential disruption at the Hotel as part of its duty to manage day-to-day hotel operations.

When an Owner wants to raise an issue at the Hotel, the path for doing so will vary depending on Manager. For large Branded Managers, there may be a regional vice-president or designated Manager representative. Even for larger independent Managers, the contact may be above the property level and part of a regional or corporate home office structure. What is most challenging for an Owner to achieve is immediate and direct contact with Executive Personnel at the Hotel. This is consistent with the concept that Manager is the sole and exclusive Operator of the Hotel and, in most instances, the employer of the Hotel Employees. By way of contrast, particularly in certain locations, with Hotels that have unique stories or market histories, or luxury amenities, Owner will negotiate for some direct lines of communication at the Hotel. One form of resolution is for Owner to be permitted to contact the Hotel's General Manager and the Hotel's director of finance/comptroller directly to discuss matters relating to the Hotel as long as the contact does not unreasonably interfere with the operation of the Hotel. It is a good practice for Owner and Manager to have a schedule of regular meetings at the Hotel. These meetings often coincide with Manager's delivery to Owner of either the monthly reports or quarterly reports for the Hotel. Although many Managers have their own standard reporting formats, the Hotel Management Agreement can specify any specific Owner requirements for the reports that Owner may require to comply with the requirements of Owner's investors or Mortgagees.

Employee benefits

Because Manager is often the employer of the Hotel Employees in U.S.-based Hotel projects, particularly with Branded Managers, Manager will include in the Management

Agreement the right to enroll the Hotel Employees in retirement, health and welfare employee benefit plans substantially similar to corresponding plans implemented in other Hotels with similar service levels managed by Manager or comparable Hotels in the area of the Hotel. These plans may be joint plans for the benefit of employees at more than one Hotel owned, leased, or managed by Manager or an Affiliate. While subject to certain limitations negotiated in the Management Agreement, employer contributions to these types of plan (including withdrawal liability or exit fees from plans) and reasonable administrative fees that Manager may expend in connection with the plans are paid for by Owner and are an Operating Expense. If the plan is a joint plan or a multi-property plan, the administrative expenses should be fairly and equitably apportioned by Manager among the properties covered by the plan. The apportionment should be based on the total costs of the administrative expenses multiplied by a fraction, the numerator of which is the total payroll expense of the Hotel, and the denominator of which is the total payroll expense of all Hotels participating in the joint plan.

Although Manager can relocate Hotel Employees and provide benefit packages to Hotel Employees, except to the extent specifically addressed in the Management Agreement, Owner remains primarily responsible for all employment-related costs and for compliance with certain laws specifically relating to employees. These Legal Requirements include compliance with the provisions of the Worker Adjustment and Retraining Notification Act (this is known as the "WARN Act") and any similar state or local laws on any disposition of the Hotel, on any termination of the Management Agreement or on the occurrence of any other event giving rise to the application of the WARN Act; multiemployer pension plan withdrawal liability under Title IV of ERISA; and severance obligations (pursuant to any union agreements or similar agreements). Financial responsibility in connection with all of these Legal Requirements, among others, is the responsibility and obligation of Owner, and Owner will usually go on to indemnify, defend and hold Manager harmless from and against any cost, expense, obligation, claim, or other liability that Manager may incur arising out of or in connection with any of these matters. With respect to any demand of withdrawal liability or exit fees, Manager and Owner will generally agree to cooperate to undertake any review, dispute resolution or other adjudication procedures that Owner determines should be undertaken, but all of this is at Owner's sole expense.

Employment laws

Because of the role Manager plays with respect to employees, even though it is at Owner's expense, the Management Agreement will establish that it is Manager that will be liable for any failure of the Hotel to comply with any federal, state, local and foreign statutes, laws, ordinances, regulations, rules, permits, judgments, orders, and decrees affecting labor union activities, civil rights or employment in the United States, including, without limitation, the Civil Rights Act of 1870, 42 U.S.C. §1981, the Civil Rights Acts of 1871, 42 U.S.C. §1983 the Fair Labor Standards Act, 29 U.S.C. §201, *et seq.*, the Civil Rights Act of 1964, 42 U.S.C. §2000e, *et seq.*, as amended, the Age Discrimination in Employment Act of 1967, 29 U.S.C. §621, *et seq.*, the Rehabilitation Act, 29 U.S.C. §701, *et seq.*, the Americans With Disabilities Act of 1990, 29 U.S.C. §706, 42 U.S.C. §12101, *et seq.*, the Employee Retirement

Income Security Act of 1974, 29 U.S.C. § 301, *et seq.*, the Equal Pay Act, 29 U.S.C. §201, *et seq.*, the National Labor Relations Act, 29 U.S.C. §151, *et seq.*, and any regulations promulgated pursuant to such statutes ("Employment Laws") with respect to Hotel Employees employed by Manager. Manager will be obligated to provide written notice to Owner of any violation or claims of violation of Employment Laws by or with respect to any Hotel Employee within a very short time after Manager receives a notice. Another consequence of Manager's role in the employment arena is that it is Manager who will be charged with the task of developing and implementing policies, procedures and programs for the Hotel to effect compliance with the Employment Laws. The Employment Policies will be not only compliant with Employment Laws but also consistent with industry standards in evidence from time to time for reputable Hotel management companies. This provides a legal standard as well as a market relevance standard.

Collective bargaining agreements

Many Hotels operate in the context of a collective bargaining agreement, whether in place when the Management Agreement was being initially negotiated or arising later. Manager is directly and daily involved with the Hotel Employees and is obligated to promptly inform Owner of any union organizing activities or campaigns, demands or petitions for union recognition, or any NLRB charges or proceedings undertaken with respect to the Hotel or the Hotel Employees. Whether Manager or Owner will control the negotiation of collective bargaining agreements and the level of approval or consent of the other party will be negotiated in the Management Agreement, but if Owner does not have the right to direct and control the negotiation of collective bargaining agreements and approve any collective bargaining agreement prior to its execution, Owner will still have meaningful input and Owner and Manager will often agree on a bargaining strategy before bargaining with the bargaining representative or representatives of Hotel Employees begins. One common way that this is expressed is that Manager is required by the Management Agreement to meet with Owner prior to commencing or participating in any collective bargaining negotiations to obtain Owner's direction with respect to the Hotel's negotiating strategy (including the strategic decision as to whether to commit the Hotel to participation in a negotiating group bound to mutual defense and acceptance of the group's collective decision) and to establish a framework of acceptable contract terms ("Framework"). Once the Framework is established, Manager, if Manager is the party controlling the negotiation of the collective bargaining agreement, consults with Owner as to the status of the collective bargaining negotiations and cannot, without the prior consent of Owner, negotiate in a manner or enter into any binding agreement that is inconsistent with the Framework.

It is possible that after the Framework is developed and approved by Owner and Manager there is nevertheless a disagreement during the course of the negotiations. Manager will usually reserve the right to refuse Owner's instructions if Manager receives written advice of independent counsel that to conform to Owner's instructions or directions would involve a substantial risk of causing Manager to commit an unfair labor practice or otherwise be violation of Legal Requirements governing labor relations. Of course, then Owner and Manager must quickly reconvene and examine the Framework and potential bargaining options.

Compensation of Manager

Management fees

Manager will negotiate a series of fees in consideration of the services Manager performs under the Management Agreement. The first fee mentioned in most Management Agreements is the Base Management Fee or Base Fee. The Base Management Fee is customarily paid on a monthly basis, in arrears for the immediately preceding Accounting Period. There is a market range of Base Management Fee from time to time, but the calculation is normally a percentage of Total Operating Revenue, Gross Revenues, Hotel Gross Revenues or similar term. Essentially, Total Operating Revenue is intended to capture all revenues and receipts of every kind derived from operating the Hotel and all departments and parts of the Hotel. This will include:

- income (from both cash and credit transactions) from rental of Rooms;
- food and beverage operations;
- telephone charges, stores, offices, exhibit or sales space of every kind;
- license, lease and concession fees and rentals, but not including gross receipts of licensees, lessees and concessionaires;
- income from vending machines;
- net income to the Hotel from parking;
- sale of merchandise;
- service charges (but not gratuities and service charges paid to Hotel Employees);
- proceeds, if any, from business interruption or other loss of income insurance.

Total Operating Revenue would not include the following:

- gratuities and service charges paid to Hotel Employees;
- federal, state or municipal excise, sales or use taxes or any other taxes collected directly from patrons or guests or included as part of the sales price of any goods or services;
- proceeds from the sale of FF&E;
- interest received or accrued with respect to the funds in the Reserve;
- any refunds, rebates, discounts and credits of a similar nature, given, paid or returned in the course of obtaining Gross Revenues or components thereof;
- insurance proceeds (other than proceeds from business interruption or other loss of income insurance);
- condemnation proceeds (other than for a temporary taking);
- any proceeds from any Sale of the Hotel or from any Mortgage;
- proceeds of any judgment or settlement not received as compensation for actual or potential loss of Gross Revenues;
- any funds contributed or supplied by Owner to the Hotel, including, without limitation, to fund Operating Expenses, Capital Improvements or the Reserve;
- advances or security deposits from Hotel guests or other Hotel users (other than forfeited deposits);
- amounts representing the value or cost of room occupancy, meals or other services provided as compensation to Hotel Employees (to the extent permitted by the approved Annual Budget) or as complementary benefits to any other persons (to the extent permitted by the approved Annual Budget);

- proceeds of collection of accounts receivable to the extent the receivable was previously included in Gross Revenues.

At the time Manager delivers to Owner each Monthly Report, Manager is authorized by Owner under the Management Agreement to pay its own Base Management Fee from the Operating Account. Manager pays itself by directly removing funds from the Operating Account. If the funds in the Operating Account are not sufficient to pay the Base Management Fee in full, then Owner must pay Manager any deficiency within a very short period of time after demand, usually 10 days or less.

The next fee commonly addressed in the Management Agreement is the Incentive Management Fee (IMF). An IMF is not present in every Management Agreement. Conceptually, the IMF should be bonus compensation for Manager for excellent economic performance of the Hotel resulting from Manager's stewardship. Because the Base Fee is generally calculated as a percentage of the top line Gross Revenues generated by the Hotel, the cost to generate the Gross Revenues is irrelevant. The better alignment of interests between Owner and Manager arises through Manager's motivation to earn a significant IMF, and this can only occur when Manager is concerned about and actively working to monitor and manage the Operating Expenses of the Hotel. The IMF is a matter of negotiation depending on the underlying management arrangement. The absence of an IMF has no intrinsic meaning, but IMF is a common feature for both Branded Managers and non-Branded Managers and the potential variations are almost endless.

One of the more common expressions of an IMF is a negotiated percentage of Net Operating Income for any Fiscal Year in excess of Owner's Priority for that Fiscal Year. In this configuration of the IMF, the parties are trying to express the notion that the Incentive Management Fee for any month during the Fiscal Year is only payable with respect to any month during or after the date that Owner has received Owner's Priority in full for that Fiscal Year. Any IMF payments made to Manager are subject to Owner's right to audit and dispute the payment once the Annual Operating Statement has been prepared and delivered. Manager may receive IMF payments monthly or quarterly, but the reconciliation of amounts paid or payable for any applicable Fiscal Year is part of the annual accounting for the Hotel. It is not unusual for the IMF to be paid once a year within a negotiated period of time after the Annual Financial Statement has been delivered to Owner. The presence of an Owner's Priority or Owner's Preferred Return is also not present in every Management Agreement. This aspect of an IMF provision is the topic of specific negotiations. The percentages or scale of percentages that are commonly key components of the IMF formula are, in large measure, a function of items included or excluded from the determination of Operating Expenses. For example, the Base Fee is often deducted from a standard Net Operating Income calculation in order to determine the IMF.

Another illustrative example might have the IMF as a percentage of Net Operating Income, but, in this example, Net Operating Income with respect to any Fiscal Year means the amount by which Gross Operating Profit properly attributable to that Fiscal Year exceeds the NOI Deductions for the same period. To calculate the Gross Operating Profit all of the Hotel's Operating Expenses have already been deducted from Gross Revenues, including the Base Fee. Then another set of deductions is applied to account for the NOI Deductions, and this is generally the sum of the following items payable with respect to that Fiscal Year: (i) Taxes; (ii) Insurance Costs; (iii) payments under leases of personal property not included in Operating Expenses; (iv) equipment rentals, ground lease payments and rents, parking space, and garage expenses; (v) expenses of Owner; and (vi) the amount to be contributed to any reserve accounts. Of course, this is just one example among many possibilities, all of which would have been carefully negotiated by Owner and Manager, and, in light of its importance, more likely at the Letter of Intent stage of the relationship and not initially in the Hotel Management Agreement itself.

The next fees commonly expressed in a Management Agreement cover the range of fees for some of the other services rendered or functions performed by Manager. These fees can include fees for the following:

- Centralized Services
- Brand Marketing and National Sales
- Reservations
- Construction Management.

Once again, when these fees are present in the Management Agreement, Manager simply pays itself from the Operating Account.

Expenses incurred by Manager on behalf of Owner

Whether the Management Agreement refers to Manager as an agent, independent contractor or provider of personal services, all debts and liabilities arising in the ordinary course of business of the Hotel and incurred in accordance with the approved Annual Budget and the Management Agreement are the obligations of Owner. Manager has no liability for any obligations merely by reason of its management, supervision and operation of the Hotel for Owner.

Because Manager is not liable for expenses properly incurred within the scope of the Management Agreement, it follows that Manager has no obligation to advance any of Manager's own funds to or for the account of Owner, and no liability on behalf of Owner unless Owner has furnished Manager with the necessary funds for Manager to discharge a liability on behalf of Owner. The Management Agreement might permit Manager, if Manager desires, and with prior written notice to Owner, to advance funds on Owner's behalf. If Manager then does advance funds, Owner must reimburse Manager, on demand, together with interest at an agreed upon interest rate.

The various fees earned by Manager under a Management Agreement are separate and apart from Manager's right to be reimbursed for its expenses under the concept of Reimbursable Expenses. Reimbursable Expenses is intended to capture things such as:

- travel, lodging, and meals;
- entertainment;
- telephone, telecopy, postage, courier, and delivery services;
- other expenses incurred by Manager which are related to Manager's performance of this Agreement.

The Reimbursable Expenses must be incurred by Manager in accordance with the approved Annual Budget, otherwise permitted by the Management Agreement or approved in advance by Owner. This is a reimbursement and not a profit center, so the Reimbursable Expenses are paid without mark-up by or profit to Manager.

Most Management Agreements will also specify certain expenses of Manager that are not Reimbursable Expenses and are not reimbursed, leaving them to be borne exclusively by Manager from its own funds, such as the following:

- all expenses, salaries, wages, or other compensation of Corporate Personnel;
- any expenses of corporate, regional, principal, or branch offices of Manager and its Affiliates or of any other Hotel owned or managed by Manager or its Affiliates;
- any travel expenses of Corporate Personnel which exceed the amounts permitted to be recovered by Manager pursuant to the approved Annual Budget or are not expressly permitted to be charged to the Hotel;
- any interest or penalty payment with respect to an imposition or lien on the Hotel or Owner by reason of the failure of Manager to make a payment required to be made by Manager under the Management Agreement when the funds are available.

In the manner consistent with the payment of other fees to Manager under the Management Agreement, after Owner's receipt of each Monthly Report from Manager, Manager pays itself the Reimbursable Expenses from the Operating Account.

Manager is best situated to monitor the capital needs of the Hotel and generally will be required to provide a monthly analysis to Owner of a 60- to 90-day rolling forecast of Working Capital reasonably necessary to satisfy the needs of the Hotel for its operation in accordance with the System Standard and the Management Agreement. If Manager projects that the funds in the Operating Accounts are less than the amount of Working Capital necessary to carry on the uninterrupted operation of the Hotel in accordance with the Management Agreement, Manager may, upon prior written notice to Owner with a short period to respond, require Owner to deposit additional Working Capital in the Operating Account. Manager's notice should specify the amount of required Working Capital, the uses of the additional Working Capital and the reasons for the shortfall in Working Capital. In determining the necessary amount of Working Capital required for the operation of the Hotel, Manager will consider the minimum amount necessary, in the exercise of reasonable business judgment, to operate the Hotel efficiently, in accordance with the provisions of the Management Agreement and the approved Annual Budget, taking into account then projected levels of business and cash flow generation, as well as expenditures made on a basis other than monthly, and taking into account the need of the Hotel, applying standard business practices, to maintain appropriate levels of Working Capital. Manager's analysis of the Hotel's cash flow generation will include an understanding of how seasonality impacts the Hotel.

Some Hotels that are located in major urban business centers may experience no seasonal fluctuations in demand. Because of the type of Hotel, such as a business limited service Hotel, located where demand drivers are relatively stable and constant, occupancy levels may also remain constant throughout the Fiscal Year. However, many Hotels and resorts experience a high season, low season, and shoulder seasons. Consider the luxury mountain resort that may only operate in the winter season or the coastal beach resort that will have very high occupancy in summer months but relatively low occupancy in winter. These seasonal variations impact occupancy, Average Daily Rate, proper management of staffing levels, proper off or low season maintenance and the general ability of Owner to make money and satisfy investors at certain times during the year. The proper management of these seasonal effects will be very important in both the development of the Operating Budget and determining the proper levels of Working Capital.

In appropriate situations, the methodology for assessing ongoing Working Capital needs can be further limited in the Management Agreement to a range based on a dollar amount or a calculation tied back to the approved Annual Budget. One common expression of this is that after Manager performs its calculation as to Working Capital needs, there could be an agreement that the amount of Working Capital would not exceed the amount determined based on some number of months of Operating Expenses based upon the approved Operating Budget. For example, while Manager may be authorized to calculate the anticipated Working Capital needs, the amount would not exceed three months of Operating Expenses under the approved Operating Budget.

No offset

Under a Management Agreement there is generally no right of offset. In fact, many Management Agreements specifically prohibit offset. All amounts payable by either party under a Management Agreement are considered made pursuant to independent covenants. The independent nature of the respective covenants of the parties supports the notion that neither party can set off any claim for damages or money due from the other party or any of its Affiliates.

Accounting, bookkeeping, and bank accounts

Books and records; reporting and reconciliation; distributions to owner

Books of control and account pertaining to operations at the Hotel are kept on the accrual basis and in accordance with Generally Accepted Accounting Principles as applied through the Uniform System, unless specifically provided otherwise in the Management Agreement.

The Uniform System offers a departmentalized system of accounting, including departmental statements of income, designed for the hospitality industry. Operating departments, which are those departments that produce Gross Revenues, include rooms, food and beverage, telecommunications, and departments to address various Hotel income producing amenities. Overhead departments include administrative and general, data processing, human resources, transportation, marketing and sales, guest entertainment, energy costs, and property operation and maintenance.

The Hotel's books and records use a December 31 Fiscal Year end, making each Fiscal Year or Operating Year under a Management Agreement a calendar year. With Managers having national or global reach, the Management Agreement should specify where the books and records are to be maintained, and if they will be maintained electronically with access to Owner through a secure portal or web site. A set of books will be held at Manager's corporate office, but should also be held at the Hotel. The calculations and payments of Gross Revenues, Operating Expenses, deposits to the Reserve, Hotel Base Management Fee, Gross Operating Profit, Fixed Expenses, asset management fees, Net Operating Income, the Brand Marketing and National Sales Contribution, Reservation Fees, the Incentive Management Fee, fees for any Centralized Services and distributions of Gross Operating Profit made with respect to each Accounting Period are accounted for cumulatively within a Fiscal Year, but not cumulatively from one Fiscal Year to the next. If the Hotel is outside the United States, the Fiscal Year may be other than a calendar year.

Managers provide Owners a detailed monthly accounting within a short period of time after the end of each Accounting Period for the immediately preceding Accounting Period. Managers usually have a standard form of Monthly Report used throughout Manager's System, and any Owner requested deviations from Manager's standard form would have to be separately negotiated. Managers will provide other information that may be specifically required under the Management Agreement or if the request is reasonable.

Some Management Agreements provide for a Quarterly Report in addition to the Monthly Report, all of which provide the foundation for the Annual Operating Statement that Manager delivers to Owner within a stated period of time after the end of each Fiscal Year. The Annual Operating Statement provides reasonable detail, in the form and in the manner reasonably required by Owner, of a summary of the operations of the Hotel for the immediately preceding Fiscal Year and a certificate from Manager's chief accounting officer or similar position, certifying that the Annual Operating Statement is true and correct. The Management Agreement will require that during a short period of time after Owner receives the Annual Operating Statement, the parties will make any adjustments, by cash payment, in the amounts paid or retained during the Fiscal Year to reconcile the calculations in the Annual Operating Statement. The period for a final reconciliation may be delayed to accommodate an audit if Owner is going to perform an audit of the Annual Operating Statement. The Annual Operating Statement will control over any preceding Monthly Reports or Quarterly Reports.

Because the delivery of Centralized Services, Brand Marketing Services and a National Sales Program can be so important to the financial success of the Hotel, some Management Agreements specify that the Annual Reports address each of those components and include the following:

- a detailed description of each Centralized Service and the corresponding costs as applied to the Hotel;
- a detailed description of the methodology utilized by Manager in allocating the cost for each Centralized Service among the participating Hotels;
- a detailed description of Brand Marketing Services and National Sales Program, and the corresponding Brand Marketing and National Sales Contribution as applied to the Hotel;
- a detailed description of the methodology utilized by Manager in allocating the Brand Marketing and National Sales Contribution among the participating Hotels.

This level of detail and specificity may not be present in every Management Agreement, but it is instructive as a reminder of the importance of these types of services and the benefit of measuring the service against the cost.

All the reports and statements are prepared by Manager and delivered to Owner. Monthly Reports and Quarterly Reports are generally not subject to audit, but if Owner wants to audit the Annual Operating Statement, Owner should have that right, to be exercised in the manner prescribed in the Management Agreement. One common configuration of the audit right begins with Owner's notice to Manager within a negotiated period of time after Owner's receipt of the Annual Operating Statement of its intention to audit. Owner must then commence the audit within a set period of time after Owner has delivered to Manager the notice of intent to audit, and complete the audit within a reasonable period after commencement, which will also be set forth in the Management

Agreement. If Owner does not undertake an audit within the time periods required by the Management Agreement, then that Annual Operating Statement is deemed conclusively accepted by Owner as being correct, and Owner has no right thereafter, except in the event of fraud or intentional misrepresentation by Manager, to question, examine or audit the Annual Statement.

If Owner's audit discloses an understatement of any amounts due to Owner, Manager must promptly pay Owner any amounts found to be due, plus interest at the interest rate set forth in the Management Agreement from the date the amounts should originally have been paid. If, however, the audit discloses that Manager has not received any amounts due to Manager, then Owner must pay Manager the amounts due, but usually without interest. Any dispute concerning the correctness of an audit is often settled by the Expert in accordance with the Dispute Resolution sections of the Management Agreement. The cost of an audit begins as an Owner expense that is not charged to the Hotel as an Operating Expense. However, if the audit reveals that Manager has understated Gross Operating Profit by an agreed upon threshold, which is usually expressed as a percentage, then Manager is obligated to pay the cost of the audit from Manager's own funds.

Manager's delivery to Owner of the Monthly Report for each Accounting Period is crucial for Manager and Owner because the remittance of the Monthly Report is one of the more common triggers for Manager to pay itself its monthly fees and for Manager to remit to Owner out of the Operating Accounts Owner's Remittance Amount. This is the payment to Owner of the amount by which the total funds then in the Operating Accounts exceeds the Working Capital to be maintained in the Operating Account pursuant and subject to the Management Agreement. If Owner believes the amounts being distributed by Manager are insufficient or otherwise not in conformity with the provisions of the Management Agreement, Owner should promptly advise Manager and the parties expeditiously meet to try to reach an accommodation as to the amount of distributions to be made. The Management Agreement should provide access to the Expert to resolve an impasse between Owner and Manager.

Accounts, expenditures

All accounts and funds derived from operation of the Hotel (other than the Reserve) are deposited by Manager in a bank account or multiple bank accounts that are for the benefit of the Hotel, but under the control of Manager. There are a variety of ways in which to handle the Operating Accounts. The financial institution might be designated by Manager and approved by Owner, designated by Owner and approved by Manager or designated by mutual agreement of Owner and Manager. It is wise to include a caveat for the possible designation of the financial institution by the Mortgagee as Mortgagees often require Owner to maintain an account with the Mortgagee. Regardless of which party designates the depository institution, the Operating Accounts are generally in the name of Owner with Manager identified as agent of Owner or otherwise authorized to act on Owner's behalf. The Operating Accounts are under the control of Manager, and some Management Agreements specifically go beyond that and identify Manager as acting as a fiduciary of Owner with respect to Owner's money. Whether the relationship between an Owner and a Manager is that of a fiduciary in general should be debated in the context of a Management Agreement. In its most basic configuration, Manager would be acting as a fiduciary for an Owner because Owner has gone beyond engaging

Manager to act on Owner's behalf pursuant to the Management Agreement to placing the utmost trust and confidence in Manager to manage Owner's money. When cast in the role of a fiduciary, Manager must demonstrate the highest degree of loyalty and care for Owner's money that has been entrusted to Manager. The legal standard to determine whether a party has discharged its duty to another as a fiduciary is assessed on a case-by-case basis in light of the facts and circumstances of the relationship. Some would argue that Manager's role as agent for Owner is sufficient, on its face, to make Manager a fiduciary, and it is inherent within the Management Agreement, but that is not universally accepted by Managers and may be effected by the laws specified as governing the interpretation of the document, leading to the idea that an Owner may want to raise with Manager and negotiate into the Management Agreement specific obligations of Manager with respect to the handling of money as a fiduciary. The fiduciary relationship would be specifically limited to the handling of money.

Withdrawals from the Operating Accounts can only be made by representatives of Manager whose signatures have been authorized and who are bonded or insured in accordance with the Management Agreement. This is another unique feature of the Hotel business and managed Hotels. It is Owner's money, but only Manager's representatives have access to it, with the exception of certain rights of Mortgagees that will arise under Owner's financing arrangements and loan documents. The fact that Manager essentially pays itself with Owner's money and controls the Hotel's accounts will lead to various insurance requirements to insure against malfeasance and crimes.

Manager will maintain at the Hotel as an operating business a reasonable amount of petty cash funds as may be determined by Manager from time to time. Whenever Manager is required to make a payment for Owner or the Hotel pursuant to the Management Agreement, the payments must be made from the Operating Accounts, the Reserve or petty cash funds, as applicable. Remember that these accounts contain Owner's money that Manager will use in connection with the proper operation of the Hotel. Manager will not be considered to be in default of Manager's obligations under the Management Agreement to the extent Manager is unable to perform any obligation due to the lack of available funds from the operation of the Hotel or as otherwise provided by Owner. Manager's obligations under the Management Agreement do not include advancing any of its own funds (whether by waiver or deferral of its management fees or otherwise) for the operation of the Hotel or incurring any liability of Owner unless Owner has first furnished Manager with funds necessary for the discharge of the liability.

Annual Budget

One of the most important components of a Management Agreement and one of the most crucial elements of the relationship of Manager and Owner concerns the process of approving the Annual Budget for the Hotel. This is where relationships are tested and significant negotiation occurs.

The process begins with Manager's delivery to Owner of a proposed annual budget. The initial questions are when is it delivered? What is the required form of the budget? How long does Owner have to review it? Is the proposed annual budget delivered for Owner's approval? Are there limitations on what an Owner has the right to approve? What happens if Owner and Manager cannot agree on the Annual Budget?

The Management Agreement will specify the delivery window for the proposed annual budget. There may be one time period for the initial annual budget and then a different time period for the annual budget during the balance of the Term. The delivery window should be close enough to the beginning of the next Fiscal Year for the budget to be accurate and meaningful, while at the same time far enough out to provide Owner sufficient review time. Sometimes Owners will ask for a preliminary budget further out, such as 90 days before the end of the current Fiscal Year, to be followed by the official preliminary budget on or before each November 1st during the Term. Under many Management Agreements with Branded Managers, the proposed annual budget is not subject to Owner's approval or is subject to Owner's approval but only with respect to certain limited components, that will specifically exclude anything related to Brand Standards. Many Managers have a standard form of Annual Budget and the Management Agreement may include the form of an annual budget as an Exhibit to the Management Agreement.

The Annual Budget may actually consist of a number of components including the Operating Budget, Capital Budget, and Marketing Plan and Budget. The Operating Budget includes Manager's forecast of Gross Revenues, Operating Expenses, Net Operating Income, Base Management Fee, Gross Operating Profits, Centralized Services to be provided and the cost thereof, departmental profits, a room rate plan, any anticipated Working Capital requirements, and any other information reasonably requested by Owner which relates to the Hotel or its operation.

The Capital Budget will include a detailed schedule of the amounts to be deposited to the Reserve and all anticipated expenditures to be made from the Reserve or otherwise proposed to be funded by Owner for non-routine or major repairs, alterations, improvements, renewals, and replacements to the Hotel, including the structure, the exterior façade, the mechanical, electrical, heating, ventilation, air conditioning, plumbing, or vertical transportation elements of the Building, and other non-routine repairs and maintenance to the Building, together with, to the extent then known, Capital Improvements forecast for the Hotel for future years, which is often between three and five Fiscal Years.

The Marketing Budget is derived from a reasonably detailed program for advertising and marketing the Hotel (the "Marketing Plan"). The Marketing Plan is likely to include proposed discount and complementary policies for *bona fide* travel agents, tourist officials and airline representatives. The Marketing Budget will include, on a line-item basis, the costs associated with the Marketing Plan. The development of the Marketing Plan and the components of it may vary greatly based on an overall analysis of the Hotel, including its location, size, segment within the industry, branding, management and target guest population. If the Hotel is Branded, there will be the additional need to identify marketing that will be handled by the Brand as part of a larger marketing campaign, and marketing that will be local in nature and handled directly by the Hotel and the local sales team.

If Owner or Mortgagee wants other information to be included in the proposed Annual Budget or any component of it, then those requests should be specifically negotiated in the Management Agreement, and most Managers will comply.

The procedure for Manager and Owner to meet and finalize the Annual Budget can take many forms. Usually, once Owner has received the proposed Annual Budget and advised Manager of any portions not acceptable to Owner, Manager arranges a meeting between Owner and a representative of Manager. This could be the Hotel General Manager or Owner might request one or more other members of the Executive Personnel or Corporate Personnel. Ideally, the meeting will take place at least 30 days prior to the commencement of the Fiscal Year to which the proposed Annual Budget relates.

Some Management Agreements stipulate a location for this type of meeting, such as at Owner's corporate office if it is in the same city as the Hotel or at the Hotel itself. Manager will generally be obligated to ensure that one or more of the senior Corporate Personnel responsible for oversight of the management of the Hotel attends the meeting. Owners generally use a meeting like this to seek clarity on Manager's Annual Budget and request explanations and additional information. Once Owner has that information it may be better able to respond to the Annual Budget. Owner would generally provide its written approval or disapproval of the Annual Budget, including the components of the Annual Budget, such as the Operating Budget, the Capital Budget, the Marketing Plan and the Marketing Budget, within a short period of time thereafter. The first strategy is very logical and it is simply that if Owner disapproves all or any portion of Manager's proposed Annual Budget, Owner will notify Manager in writing, including a written explanation of Owner's disapproval, and the parties will use reasonable efforts to agree on the items to which Owner disapproves. What if Manager and Owner are unable to reach agreement before the commencement of the Fiscal Year to which the proposed Annual Budget relates?

- Manager may operate the Hotel in accordance with those line items in any of the budgets to which Manager and Owner have agreed;
- as to those line items in the budgets to which Owner and Manager have not agreed, Manager may operate the Hotel in accordance with those line items as set forth in the approved Annual Budget applicable to the immediately preceding Fiscal Year as increased by the greater of pre-negotiated percentage increase in all expense line items or the actual cost of any non-discretionary expense line item;
- Manager shall make only those expenditures for Capital Improvements specifically authorized by Owner;
- either Owner or Manager can submit the disputed line items to Expert resolution.

Once there is an approved Annual Budget, either through the agreement of the parties or through the assistance of the Expert, Manager is obligated to use commercially reasonable efforts to adhere to the approved Annual Budget. This obligation is not unlimited. Manager's compliance obligation is subject to the following:

- Manager is not required to adhere to the Annual Budget to the extent that any expenses, such as the cost of utilities, insurance, the costs of labor, material, services and supplies, the occurrence of a casualty involving the Hotel, expenses incurred as a result of Legal Requirements, or other expenses, are outside of the control of Manager;
- Manager may exceed the amount of any single line item in the Operating Budget or the aggregate amount of the Operating Budget by up to a negotiated percentage, so long as the excess expenditures are detailed in the applicable Monthly Report, including the specific reasons for the excess expenditures;
- Manager may exceed the expenses set forth in the approved Annual Budget to the extent required to pay any expenses which, in Manager's good faith judgment, are immediately necessary to avoid or minimize any actual or potential injury to persons or damage to the Hotel or other property, and any imminent risk of criminal or civil liability of Manager and its Affiliates; provided, however, that Manager will usually be obligated to first make a reasonable attempt to obtain Owner's approval of those

expenditures, by both telephonic and electronic communication, and in no event would those expenditures exceed a pre-negotiated cap, without the prior approval of Owner.

Owner and Manager can always revise an approved Annual Budget in the same manner as an Annual Budget is initially approved in order to maintain or improve expectations of the Annual Budget as originally approved.

Where Owners and Managers spend most of their time in negotiating the budget approval provisions of a Management Agreement, it is the scope of Owner's right to approve the Annual Budget. This is particularly apparent when Manager is a large Branded Manager with specific System Standards or Brand Standards that each Hotel in the system is required to adhere to.

Repairs, maintenance, and replacements

Repairs and maintenance to be paid from gross revenues

Under the Management Agreement, Manager has the obligation and duty to maintain the Hotel in good repair and condition, in conformity with the System Standards, and in compliance with all Legal Requirements. As part of the discharge of these obligations, Manager will make all routine maintenance, repairs, and minor alterations that Manager reasonably determines are necessary. "Routine maintenance, repairs, and minor alterations" is intended to address matters that are normally booked as an expense under Generally Accepted Accounting Principles. The cost associated with this type of maintenance, repairs, and alterations is paid from Gross Revenues and is treated as an Operating Expense in determining Gross Operating Profit. The Reserve maintained by Manager for the Hotel is not used for the purpose of paying the cost of this type of repair and maintenance.

Repairs, maintenance, and equipment replacements to be paid from reserve

The Reserve described earlier is maintained by Manager, usually in a bank designated by Owner, and in the name of Owner. As with the Hotel's Operating Accounts, the Reserve is under the control of Manager, even though in all other respects it is the property of Owner. Withdrawals from the Reserve are made solely by representatives of Manager, and those Manager Representatives will be authorized and bonded or insured under the Management Agreement. The use of the funds in the Reserve are subject to the Annual Budget, but generally are used to pay the cost of:

- replacements and renewals related to the FF&E at the Hotel;
- routine or non-major repairs and maintenance to the Building which are normally capitalized (as opposed to expensed) under Generally Accepted Accounting Principles, such as exterior and interior repainting and resurfacing building walls, floors, roofs, and parking areas.

Once the Hotel opens for business to the public on a paying basis, Manager will take from Gross Revenues and make a monthly deposit to the Reserve. The amount of that monthly deposit is generally the greater of (i) any amounts required by any Mortgage, or (ii) a negotiated schedule for deposits to the Reserve that might look something like this:

- through the first full Fiscal Year, an amount equal to 1% of gross revenues.
- for the second full fiscal year, an amount equal to 2% of gross revenues.
- for the third full Fiscal Year and for each Fiscal Year thereafter during the Term, an amount equal to 3% of Gross Revenues.

Any schedule of Reserves will remain subject to requirements imposed by the Mortgagee. Those requirements may go beyond simply the amount of the monthly deposit and include the institution where the Reserves will be held and the mechanics of the account, such as a variety of lockbox and cash management obligations.

The funds to be deposited into the Reserve are deducted from Gross Revenue and deposited by Manager. They do not pass through the hands of Owner, even though those funds are technically Owner's money.

Under the Management Agreement, Manager will use the Reserve for replacements and renewals to the FF&E of the Hotel, and repairs to the Building. The use of the Reserve and the determination of the amount of the monthly deposit to the Reserve is an estimate of what may be needed, not a guaranty. Owner will acknowledge in the Management Agreement that setting aside funds in the Reserve does not guaranty that the Reserve will provide sufficient funds to meet Owner's obligations under the Management Agreement. The funds in the Reserve generally do not leave the Reserve. At the end of each Fiscal Year, any amounts remaining in the Reserve are carried forward to the next Fiscal Year. Proceeds from the sale of FF&E no longer necessary to the operation of the Hotel are also added to and retained in the Reserve. Both proceeds from the disposition of FF&E and interest earned on amounts held in the Reserve can be used to reduce the required contributions to the Reserve but that is not a given without negotiation. If at any time during the Fiscal Year the amount on deposit in the Reserve is insufficient for replacements and renewals to the FF&E of the Hotel, and repairs to the Building, Manager will seek additional funds from Owner, and Owner will need to comply or risk being in default under the Management Agreement. This is more than an academic observation. Over the term of a Management Agreement and the useful life of a Hotel and all of its component parts, it is unlikely that the amount on deposit in the Reserve will be sufficient to pay for all that will be necessary to maintain a fresh and competitive Hotel. The FF&E Reserve may be called upon to fund the replacement of aging or obsolete technology, such as televisions and telephones, the replacement of floor and wall coverings, and to undertake basic routine interior and exterior maintenance. However, it may also be needed to fund modifications in Brand Standards, such as logo and signage changes, or improvements dictated by changes in legal requirements, such as recent changes to the Americans with Disabilities Act that lead to the placement of pool lifts at almost every hotel pool and spa.

Insurance, damage, and condemnation

Insurance

In developing the management agreement, there are five main topics to address in regards of insurance as shown in Table 3.1.

In the context of a large Branded management company, unless the Management Agreement provides otherwise, it is Manager that will obtain the insurance for Hotel operations. The procurement process can also be segregated with Owner taking control

Table 3.1 Topics to address in insurance claims

Item	Description
Risk of loss	The party that holds responsibility (financial and administrative) in the event of a loss
Insurance	The party that holds responsibility to maintain the insurance in force
Indemnification	Outlines the provisions for one party indemnifying the other in the event of a financial loss
Hold harmless	Linked to risk of loss and indemnification, as above. Essentially, one party holds the other harmless from risk of loss
Waiver of subrogation	A waiver of subrogation essentially eliminates the potential for one party's insurer to seek subrogation recovery for the other party

of the elements of insurance that are related to the ownership of the land and improvements rather than the operation of the Hotel.

Some of the insurance coverages (not the amounts, as those will vary due to a number of circumstances) that Manager will typically obtain in Manager's name and also insuring Owner's risk of loss (to the extent available and at commercially reasonable rates) include:

- commercial general liability insurance (including broad form endorsement and coverage against liability arising out of the ownership or operation of motor vehicles), not eliminating cross-liability and containing a severability of interest clause;
- liquor liability;
- worker's compensation insurance or insurance required by similar employee benefit acts;
- fidelity;
- employee crime;
- employment Practices Liability;
- errors and Omissions;
- umbrella excess liability;
- Business Real and Personal Property and Business Interruption;
- advertiser's liability;
- broad form money and securities;
- blanket oral and written contractual liability;
- wrongful dismissal;
- Owner's protection liability;
- garage keeper's excess liability;
- valuable papers.

If Owner decides not to have Manager provide all or any portion of the insurance either listed here or appropriate for the Hotel, then Owner and Manager will need to contractually outline the responsibility for insurance. Generally, the insurance for operational coverage for the Hotel is paid for by the Hotel, often, but not always, as an Operating Expense, and the insurance for first party damage-related coverage for the Building is an

Owner's expense. Manager will certainly have other insurance for the protection of Manager at a corporate level or at the property level that will be paid from Manager's own funds.

Insurance that Owner is likely to obtain and hold in its name include the following:

- insurance covering the Building, the Installations and the FF&E (including boiler and machinery insurance, and alterations, rebuilding and addition coverage but excluding damage resulting from earthquake, war, and nuclear energy);
- business interruption insurance covering loss of income resulting from interruption of business resulting from physical damage caused by the occurrence of any of the risks affecting the Hotel. This loss of income would also cover the management fees due to Owner in the event of a property-related loss event at the Hotel, such as a fire.

Business interruption insurance can also be independently obtained by Manager, at Manager's sole expense, to provide independent coverage for Manager's loss of income/management fees as a result of a covered event. All Management Agreements should outline the party that will obtain the business interruption insurance and the covered period of interruption. Of course, Owner, as an expense of Owner, can seek any other insurance coverage that it deems prudent.

Regardless of the party that procures the insurance, the cost of the bulk of the insurance premiums will be an Operating Expense. Some ownership-related insurance may be a Fixed Expense and not an Operating Expense, but that is a matter of negotiation that also has implications for the IMF and Owner's Preferred Return.

Policy requirements

The insurance policies obtained by Manager will name Manager as the insured party, together with other insured parties like Owner and a Mortgagee. If Owner obtains insurance outside of Manager's insurance program made available at the Hotel, that insurance must name Manager as an additional insured party. All liability insurance policies obtained by Owner will name Owner as the insured party and should also name as additional insureds Manager and any other parties required to be named by the terms of any Mortgage. The named insured in an insurance policy is the party or parties specifically designated by name as an insured(s) in the insurance policy. There are other parties that, although unnamed, fall within the policy definition of an "insured," provided the policy contains a Broad Named Insured wording extension.

A loss payee is a party that is entitled to all or part of the insurance proceeds in connection with the covered property in which it has an interest. An additional insured is a party that is not automatically included as an insured under the insurance policy, but who is included or added as an insured under the policy at the request of the named insured. A named insured may provide additional insured status to others out of a desire (and/or if contractually required) to protect the other party because of a close relationship with that party or to comply with a contractual agreement requiring the named insured to do so. In liability insurance, additional insured status is commonly used in conjunction with an indemnity agreement between the named insured (the "Indemnitor") and the party requesting additional insured status (the "Indemnitee"). Having the rights of an insured under its indemnitor's commercial general liability policy is viewed by most indemnitees as a way of securing and supporting the promise of indemnification. If the indemnity agreement proves unenforceable for some reason, the indemnitee may still be able to obtain coverage for its liability by making a claim directly as an additional insured under the indemnitor's policy.

An additional named insured is a party, other than the first named insured, identified as an insured in the policy declarations or an addendum to the policy declarations with the status of named insured. This party would have the same rights and responsibilities as a named insured in the policy declarations.

An *additional named insured* can be contrasted with an *additional insured*, in that an additional insured party is added to a policy as an insured but not as a named insured. In principle, only the named insured has rights to request policy changes and cancel the policy. The additional insured is essentially an insured party, on the basis of ATIMA (as their interests may appear).

The Management Agreement will go on to address the form of policy itself as well as criteria for the insurance companies who can provide the insurance.

If one party is obtaining the insurance, the other party usually has the right to reasonably approve the form and the company. Of course, the insurance requirements must also be consistent with the requirements of any Mortgage. For larger management companies using blanket policies, which is permitted by most Owners and Mortgagees, so long as the portion of the premium for any blanket or master policy that is allocated to the Hotel as an Operating Expense is determined in an equitable manner by Manager, subject to verification by Owner.

All insurance policies must specify that they cannot be canceled or modified without prior written notice to Owner, and usually with not less than 30 days prior notice, although that time period may be longer due to the requirements of a Mortgagee. It is also standard for the insurance policies to provide, unless it is not customarily obtainable from the insurance company providing the insurance, that the insurance company will have no right of subrogation against Owner, Manager, any party to a Mortgage or any of their respective agents, employees, partners, members, officers, directors or beneficial Owners. If these parties are actually named insureds under the policy, there will also be a specific waiver of subrogation.

All insurance policies for the Hotel must be issued as a primary policy and not "excess over" or contributory with any other valid, existing and applicable insurance in force or on behalf of Owner. The Management Agreement will also require that the insurance be issued by insurance companies of good reputation and of sound financial responsibility with a minimum "Best" rating by two Insurance Reports, or as otherwise required by any Mortgage.

If any construction, renovation or other similar work will occur at the Hotel, Manager or Owner will be required to include insurance requirements in each contract. There will be some variation, but generally each contractor will be required to carry these types of insurance:

- workers' compensation;
- employers' liability;
- commercial general liability;
- professional liability for architects, engineers and other professionals.

Damage and repair; casualty and condemnation

Hotel properties can suffer a partial or total casualty or be the subject of a condemnation proceeding just like any other asset class. What happens in those situations will be negotiated between the parties, but there are some overarching considerations that the Management Agreement will address.

The first action item concerns notice. With Manager being onsite and solely responsible for the day-to-day operations of the Hotel, Manager will have a duty to promptly notify Owner if the Hotel is damaged by fire or other casualty. The next item is the evaluation process. Manager must cooperate with and assist Owner in the process of evaluating the extent of the damage, filing and pursuing insurance claims, developing plans for the restoration of the Hotel and modifying the approved Annual Budget to reflect the effect of the casualty on operations, subject, at all times, to the role of the Mortgagee with respect to how and when insurance proceeds might be used by Owner. All of this should be collaborative, but require Owner's approval in its commercially reasonable discretion. The Management Agreement does not automatically terminate upon a casualty. Owner may have a right to terminate the Management Agreement in certain circumstances, but until the effective date of that termination the Management Agreement remains in full force and effect subsequent to such casualty. The conditions and thresholds that may give rise to the right to terminate a Management Agreement following a casualty or condemnation will be specifically negotiated by the parties, but if Owner has, and then exercises, a right to terminate the Management Agreement, there is usually a "spring-back" period that will cause the terminated Management Agreement to spring back to life should Owner later decide to rebuild the Hotel or use the property as a Hotel. If the Management Agreement is terminated and then reinstated, the remaining Term will be the Term that was remaining under the Management Agreement as of the date of its earlier termination. If Owner elects to restore the Hotel, there will be a period of time during which Owner must use commercially reasonable efforts to complete the restoration, subject to a Manager's right to terminate the Management Agreement if the restoration process is not completed on time.

One point that is often debated is whether or not Manager will be entitled to a termination fee from Owner if Owner elects not to rebuild the Hotel after a casualty when insurance proceeds are available to Owner. From Manager's point of view, Manager is being deprived of the benefit of its bargain as a result of the casualty and Owner's determination not to rebuild. Owner will collect on its insurance while Manager will lose the remaining Term of the Management Agreement and the management fees it would have earned under the Management Agreement, subject, perhaps, to Manager's receipt of business interruption insurance if such a policy was in place for the benefit of Manager. From Owner's point of view, Owner is the party suffering the real loss, and the determination of whether or not to rebuild will depend upon a number of factors beyond the casualty itself, such as access to insurance proceeds, the adequacy of those proceeds, Mortgagee requirements and the possible impact of intervening changes to zoning and building codes. This is an element of the Management Agreement that will be subject to negotiation with potentially differing outcomes.

Condemnation proceedings create similar concerns. Manager's initial duties under the Management Agreement will be to promptly notify Owner within a very short period of time after Manager learns of a condemnation proceeding, and to cooperate with and assist Owner. An Owner may require or seek differing levels of assistance, but the initial areas of cooperation typically involve the following:

- evaluating the effect of the condemnation;
- pursuing any negotiation, litigation, administrative action or appeal relating to the condemnation award;
- developing plans for the restoration of the Hotel;
- modifying plans and budgets to account for the effect of the condemnation on Hotel operations.

Owner will have the right to terminate the Management Agreement upon notice to Manager if all or a substantial part of the Hotel is taken through condemnation. If the condemnation is a partial taking that does not render it impractical or unwise to operate the remainder of the Hotel in accordance with the System Standards, then the Management Agreement would not be terminated. Owner would undertake alterations or modifications to the Hotel as reasonably necessary for the Hotel to continue to operate as a Hotel of the type and class immediately preceding the condemnation.

Condemnation is controlled by state law, which will cause there to be variations in the approach under the Management Agreement based on where the Hotel is located. That being said, more often than not the Management Agreement will provide that any condemnation award or compensation belongs to Owner, but Manager has the right to bring a separate proceeding against the condemning authority for any damages and expenses specifically incurred by Manager as a result of the condemnation, as long as any award to Manager is in addition to and does not reduce the award payable to Owner. Although sometimes seen in a Management Agreement, it is much less common for Manager to be entitled to a termination fee from Owner if Owner elects to terminate the Management Agreement after condemnation proceedings that render it impractical to continue or resume Hotel operations.

Taxes

Imposition

Owner is responsible for paying or for providing Manager the funds necessary to pay taxes and impositions assessed against the Hotel. In this context, "Impositions" is intended to cover all real estate and personal property taxes, levies, assessments, association fees and charges, and similar charges on or relating to the Hotel attributable to the Hotel or any of its component parts.

The administrative task of actually rendering payment to the taxing authority may be delegated to Manager under the Management Agreement, but Owner has the obligation to make funds available to make the payments.

So long as funds are available either in the Hotel Operating Accounts or as provided by Owner, Manager will not only make the payment in a timely fashion, but will furnish Owner, before the respective dates on which Impositions will become delinquent, proof of payment. Owner will retain the right to contest the validity or the amount of the Impositions, but Owner must exercise that right in a manner that does not jeopardize Manager's management of the Hotel. For its part, Manager will cooperate with Owner and execute documents, pleadings or other instruments needed for Owner to contest the Impositions.

The prosecution of tax assessment appeals should be specifically addressed in the Management Agreement. This is often either overlooked or given inadequate attention only to become a point of contention later. Whether Owner or Manager will control this process is important. Paying Impositions impacts both Manager and Owner, and achieving a tax savings if the assessment of the Hotel is deemed inaccurate will impact the distributable cash available in the Operating Account. Generally, tax assessment appeal cases are handled on a contingent fee basis by lawyers and other professionals knowledgeable in this area. However, this type of fee structure is not always the most appropriate and can result in a diminished return for the Hotel. The fee structure should not be assumed, but rather should be examined and negotiated on a case-by-case basis. Either Manager or Owner will probably be able to discharge the responsibility, but both sides should have confidence that it will be thoughtfully discharged. The allocation of responsibilities and oversight should be set forth in the Management Agreement.

Fixed expenses

Owner may also direct Manager to pay certain Fixed Expenses out of the Hotel's Operating Accounts or with funds Owner makes available to Manager. Under those circumstances, it is incumbent on Owner to provide Manager with complete and timely notice of all Fixed Expenses Manager is to pay, and then Manager would pay the Fixed Expenses before they become delinquent and provide proof of payment to Owner.

Representations, warranties, and covenants of owner

It is customary for Owner to make a number of representations, warranties and covenants in the Management Agreement. If Owner is not truthful or any representation, warranty or covenant is breached by Owner, there is a default under the Management Agreement.

The representations, warranties and covenants of Owner will be customized and fashioned to meet the needs of each Management Agreement transaction between an Owner and a Manager. Some of the more typical topics that should be addressed as to Owner include the following:

- Owner's right to possess the land. That Owner owns fee simple title to the land or Site or has rights pursuant to a ground lease or other arrangement;
- Owner will, throughout the Term, maintain the title to the Site, and following the Opening Date, to the Hotel, free and clear of any liens, charges and encumbrances of any nature or kind other than any encumbrances permitted under the Management Agreement;
- from and after the Opening Date, Owner will not act or fail to act in such a manner that would cause the Hotel to become subject to any zoning, land use, or development control restrictions of any nature or kind that prevent, impede, or delay the furnishing, equipping, marketing, maintenance, operation, management, supervision, or direction of the Hotel as contemplated by the Management Agreement;
- from and after the Opening Date, except as provided in an approved Annual Budget, Owner will refrain from making an agreement or contract for any alteration, addition, improvement, design or other modification to the Hotel that could reasonably

be expected to have an adverse material effect on the financial performance of the Hotel;

- Owner has no actual knowledge of there being Hazardous Materials on any portion of the Site and no Hazardous Materials have been released or discharged on the Site (which may, particularly in the case of new construction or a material renovation and rehabilitation, be supported by a Phase I Environmental Site Assessment);
- if any Hazardous Materials are found on the Site, Owner and Manager will comply with all Legal Requirements relating to the abatement, remediation or cure. All costs and expenses relating to any abatement or removal of Hazardous Materials found on the Site or the Hotel, and of compliance with Legal Requirements, are the expense of Owner from Owner's own funds, neither as an Operating Expense nor from the Reserve, and shall be treated as an expenditure by Owner; but on an ongoing basis, compliance with Legal Requirements, such as approvals for use and storage, will be an Operating Expense.

Representations, warranties, and covenants of manager

Manager will also make representations, warranties and covenants in the Management Agreement. Just like Owner, if Manager is not truthful or any representation, warranty or covenant is breached by Manager, there is a default under the Management Agreement.

The representations, warranties and covenants of Manager will be customized and fashioned to meet the needs of each Management Agreement transaction between an Owner and a Manager. Manager makes fewer representations than Owner, but a few examples of typical topics that should be addressed as to Manager include the following:

- all existing and future Hotels in the System will be managed by Manager or an Affiliate of Manager;
- Manager possesses the requisite knowledge, skill and expertise to properly discharge its duties under the Management Agreement.

Management of all Hotels in the System will be subject to exceptions because it is not applicable in the context of a franchise and there are some situations where the Hotel is part of a collection or similar structure where the elements of the Brand have been licensed to Owner, but the Hotel is managed by a third-party Manager.

Mutual representations, warranties, and covenants

Each Owner and Manager may make other representations, warranties and covenants to the other party such as the following:

- each party will provide to the other (generally further expanded to include promptly or within a pre-negotiated period of time) a copy of any notice in writing received from any governmental authority or Mortgagee. These notices might be advising Owner or Manager of:
  - defects in the construction, state of repair or state of completion of the Hotel;
  - ordering or directing that any alteration, repair, improvement or other work be performed at the Hotel;

- non-compliance with any building permit or Legal Requirements;
- any threatened or impending condemnation or taking of property;
- any notice of non-compliance, control order, stop order, or any other order or decree made by any governmental authority with respect to any environmental issue or under any environmental protection legislation.

- from and after the Opening Date, each party will provide to the other the full particulars of any plans to widen, modify, or realign any of the streets or other public thoroughfares or rights-of-way contiguous to the Hotel or any development controls, planning decisions or other such regulations that might in any way restrict the operation of the Hotel;
- each party will provide to the other written notice of any action, suit or proceeding of which either party becomes aware which is pending or, to the knowledge of either party, threatened at law or at equity or before or by any governmental authority that affects or may affect the Hotel or the Management Agreement;
- each party will take, or refrain from taking, actions to ensure that the party is not in default under or in breach of any material contract, agreement, or other instrument beyond any applicable notice and cure period;
- each party is each duly organized, validly existing, and in good standing in the state of its formation and is qualified to do business in the jurisdiction in which the Hotel is located;
- each party will maintain its good standing in the jurisdiction of its formation and its due qualification to do business and good standing under the laws of the jurisdiction in which the Hotel is located;
- each party has taken all steps necessary to be authorized to enter into the Management Agreement and to perform its obligations thereunder;
- each party makes the appropriate representations, warranties and covenants as required by the U.S. Department of Treasury's Office of Foreign Asset Control, Section 1 of U.S. Executive Order 13224 issued on September 23, 2001, and all other Legal Requirements.

Subordination

The subordination of the Management Agreement to the lien of Owner's Mortgages can take many forms and configurations. Often, the perceived relative strength of Manager, as a large national Brand as compared to a smaller Brand or an independent third-party Manager, can be determinative of the outcome, particularly with respect to a Manager requiring non-disturbance from the Mortgagee. A Manager that can achieve a Subordination, Non-Disturbance, and Attornment Agreement will not be subject to termination by the Mortgagee following a borrower default under the Mortgage and the Mortgagee's enforcement of its rights and remedies pursuant to the loan documents through foreclosure or a deed in lieu of Foreclosure. In the event that the Mortgagee acts to cure Owner's default under the Management Agreement or performs the obligations specified in the SNDA, the Mortgagee will often gain certain other special Mortgagee rights under the Management Agreement and the SNDA intended to accommodate a Mortgagee that involuntarily steps into the shoes of Owner.

Manager will be asked to acknowledge and agree that Manager's rights under the Management Agreement are junior and subordinate to the lien of any and all Mortgages encumbering the Hotel, whether now or hereafter existing. Manager may also be asked to indicate that the Management Agreement is a contract for services which does not create any interest for Manager in the Hotel or a lien or encumbrance on the Hotel of any kind and does not run with the land.

The subordination provisions of a Management Agreement often mention or incorporate many related topics, such as Owner's right to encumber the Hotel with the Mortgage in the first instance. Owners will negotiate for an acknowledgment that Owner is permitted to encumber the Hotel in any amount and on the terms Owner may determine, in its sole discretion and without Manager's approval. This will be contrary to some Management Agreement forms that stipulate various limitations on Owner's financing and Manager's prior approval of certain financing, such as limitations based on loan-to-value, loan-to-cost, and debt service coverage. For a Manager, it is important that Owner not overleverage the Hotel and leave Manager concerned that Owner may lack the financial strength to provide cash to the Hotel if Gross Revenues do not support the Operating Expenses and Owner's debt service. Owner, on the other hand, wants the maximum flexibility achievable as to the financing of the Hotel.

Because non-disturbance is so important to Managers, many Management Agreement forms start with a provision that obligates Owner to deliver to Manager a subordination and non-disturbance agreement, in form and content reasonably acceptable to Manager, from each Mortgagee. If this is an absolute covenant and obligation of Owner, then Owner's failure to cause its Mortgagee to agree to enter into such an agreement constitutes an Owner's event of default under the Management Agreement, which can lead to Manager's right to terminate the Management Agreement and possibly seek damages from Owner, and may also preclude Owner from obtaining favorable financing terms. Many Owners and Managers do not want the Management Agreement to represent an agreement by either party to accept a certain position with respect to a Mortgagee's subordination agreement. A common middle ground is for the Management Agreement to obligate Owner and Manager to negotiate in good faith with any Mortgagee with respect to the terms and provisions of the Subordination, Non-Disturbance, and Attornment Agreement and reasonably consider any revisions or modifications proposed by Mortgagee or counsel for Mortgagee. There is no rule of general applicability or rule of thumb for what happens when the negotiation of the SNDA fails to achieve a mutually satisfactory result. If a general rule had to be stated, it would be that failure is not an acceptable option, and the parties acknowledge that they simply must continue to negotiate and deliberate until an acceptable solution is achieved.

Similarly, any provisions of the Management Agreement regarding cash management or the handling of Hotel funds are subject to any requirements regarding cash management and handling of Gross Revenues and other Hotel funds, including but not limited to lockbox or similar arrangements, of the Mortgagee. Here again, Owner will seek Manager's commitment to join in any agreement, acknowledgment, or consent with respect to subordination and non-disturbance reasonably required by the Mortgagee, so long as the arrangements with Mortgagee do not materially and adversely affect Manager's rights pursuant to the Management Agreement to receive funds that Owner is obligated to provide pursuant to the Management Agreement.

Remember to also refer to Chapter 6 for further explanation of Subordination-related issues.

Events of Default

The Management Agreement will include a schedule of occurrences that would constitute an "Event of Default" under the Management Agreement. A common schedule of defaults applicable to both Owner and Manager would include the following:

- the filing of a voluntary petition in bankruptcy or insolvency or a petition for reorganization under any bankruptcy law, or the admission in writing by a party that it is unable to pay its debts as they become due;
- the consent to an involuntary petition in bankruptcy or the failure to vacate, within a stated period of time from the date of entry any order approving an involuntary petition;
- the entering of an order, judgment or decree by any court of competent jurisdiction, on the application of a creditor, adjudicating either party as bankrupt or insolvent or approving a petition seeking reorganization or appointing a receiver, trustee, or liquidator of all or a substantial part of the party's assets, and the order, judgment, or decree's continuing unstayed and in effect for an aggregate period of time;
- the failure to make any payment required to be made in accordance with the terms of the Management Agreement, as of the due date as specified in the Management Agreement, and the failure of the party to cure the payment failure within 10 days after receipt of written notice from the non-defaulting party demanding cure;
- either party or any of their respective Affiliates is or becomes a Specially Designated National or Blocked Person;
- the failure of either party to perform, keep, or fulfill any of the other covenants, undertakings, obligations or conditions set forth in the Management Agreement, and the continuance of that failure for a period of time, such as 30 days after the defaulting party receives written notice from the non-defaulting party of the failure, or, if the failure cannot reasonably be cured within the stated time period, if the defaulting party fails to commence the cure of the failure within the stated time period or thereafter fails to diligently pursue efforts to completion, but in any event the failure must be cured within a stated period of time following the defaulting party's receipt of written notice;
- if the general Manager or the director of finance/comptroller of the Hotel commits willful misconduct, fraud, or theft against Owner or the Hotel, but only if and to the extent that Manager fails, within a stated period of time after Owner's written request, (i) to hold Owner harmless from all losses resulting from the willful misconduct, fraud or theft, and (ii) to the extent the willful misconduct, fraud or theft is material or has become a repetitive action by the general Manager or director of finance/comptroller of the Hotel, to terminate the employment of the general Manager or director of finance/comptroller of the Hotel;
- if either party effects a Transfer in violation of the transfer provisions of the Management Agreement.

An important negotiation may occur with respect to the expanded or limited use of the fiduciary concept that was initially raised in connection with the hotel's accounts and expenditures. Some managers assert that Manager is not a fiduciary for Owner. Some Owners negotiate for specific provisions with limited applicability making Manager the fiduciary with respect to the handling of money. When a Management Agreement includes language to the effect that Manager will discharge its obligations under the Management Agreement and manage the Hotel to maximize the return to Owner, it generally should be read and understood in the context of the Management Agreement as a whole, including all of the provisions that express that all budgets have been prepared in good faith, but are merely estimates of performance, and that no specific levels of return are assured. These concepts will influence the negotiation when a strong Owner seeks to specifically make Manager a fiduciary generally under the Management Agreement. A court hearing an Owner and Manager dispute may use fiduciary language in a decision of a dispute, but some Managers may assert that the fee structure of a typical Hotel Management Agreement does not justify Manager assuming the role of a fiduciary for Owner and such a relationship is not contemplated by the parties. Regardless of the positions asserted with respect to the fiduciary concept, it is generally not a Manager event of default for Owner to fail to realize a return on its investment in the Hotel. Once again, this is carefully negotiated and will be played out in different forms in the concepts of Owner's Preferred Return and the Performance Test.

Remedies

If an Event of Default occurs under the Management Agreement and is not cured, it gives rise to a number of possible courses of action by the non-defaulting party. Remember that the non-defaulting party is not obligated to do anything if there is an Event of Default. One possible remedy for the non-defaulting party is to do nothing, and in the Hotel operations and management environment, there can be adequate commercially sound reasons for either Owner or Manager to take no action in the face of an event of Default by the other party. Should the non-defaulting party desire to act upon an event of Default, some of the rights that are customarily available to pursue under the Management Agreement are as follows:

- terminate the Management Agreement, and, if by Owner following a Manager is Event of Default, without payment of a Termination Fee;
- institute any proceedings permitted by law or equity (provided they are not specifically barred under the terms of the Management Agreement);
- avail itself of any remedies specifically included within the four corners of the Management Agreement.

When the Event of Default concerns the payment of money, the Management Agreement will stipulate that interest will accrue at the agreed upon interest rate so long as the amount remains unpaid.

Because Management Agreements often involve rights and obligations for which the appropriate remedy is not the payment of money from one party to the other, there will be some operative provisions and covenants for which the most appropriate remedy is injunctive relief and other similar equitable remedies. Whether the remedies stated in the Management Agreement are intended to be the sole remedies available to the parties or just part of the rights and remedies available to the non-defaulting party should be included in the Management Agreement. Similarly, the Management Agreement will specify if and the extent to which the parties have agreed to limit their remedies to dispute resolution through mediation or arbitration, and the limitations, if any, upon use of those special remedies.

Performance termination

A common, but specially crafted, Owner remedy is an Owner's right to terminate the Management Agreement, without the payment of a termination fee or other fee or expense (other than Manager's compensation set forth in the Management Agreement accrued through the effective date of Termination), due to Manager's failure of a performance test as expressed in the Management Agreement.

Performance Tests are initially designed and devised by Manager, but the final result should reflect an Owner and Manager negotiation, ideally during the development of the Letter of Intent. There are significant variations as to the elements of the test. Nevertheless, one of the more common variations of the Performance Test possesses the following elements:

- it must be failed for two consecutive performance test periods;
- it will be a two-pronged test and Hotel operations must fail both prongs of the test;
- if Owner wants to terminate Manager based on a Performance Test failure, Owner must provide a termination notice to Manager within a stated period of time, such as 60 days, after Owner's receipt of the Performance Test Financial Statement for the second performance test period;
- there will be a stabilization period during which the Performance Test will not apply (generally the initial two to five years of the Term);
- Manager will have a limited number of opportunities to make a cure payment to Owner to eliminate the shortfall that represented the Performance Test failure, which can be curing the last year of the failure or curing both years of the failure.

The most common tests are a pre-negotiated percentage of the (i) amount of the Hotel Gross Operating Profit set forth in the approved Annual Budget for each Performance Test Period, and (b) RevPAR Index of the Competitive Set for each Performance Test Period. The effective date of Termination is stated in the Performance Termination Notice but usually does not occur earlier than 60 days from the delivery of the Performance Termination Notice.

The Management Agreement will further limit Owner's right to seek termination of Manager if the Performance Failure is a result of any of the following:

- an Extraordinary Event (a defined term that will also be negotiated because of its impact on the application of the Performance Test);
- a material reduction in occupancy resulting from an Extraordinary Event;
- a taking by eminent domain of all or a portion of the Hotel;
- an Event of Default of Owner under the Management Agreement, such as failing to provide Working Capital;
- a casualty;
- a major renovation of the Hotel or any other facilities or venues at the Hotel.

The other tool available to a Manager is to add to the Management Agreement the ability to negate a Performance Termination Notice by making a cure payment to Owner. The calculation of the amount of the cure payment and any limitation on the number of times a Manager can make a cure payment to negate a performance test failure will be as specifically negotiated between Owner and Manager. Owners and Managers may negotiate variations in how the cure payment will be applied. Examples include Manager paying to Owner or as directed by Owner, from Manager's own funds, an amount equal to the difference between actual Gross Operating Profit achieved for the second performance Test Period and 90% of the Gross Operating Profit in the approved Annual Budget for the Performance Test Period, or Manager paying the necessary amount for both Performance Test Periods. A further variation is for the next occurring Fiscal Year to constitute the second rather than the first Performance Test Period of a new Performance Test Period, making that third consecutive year of non-conformance with the Performance Test another failed Performance Test.

Termination on sale of the hotel

The ability of an Owner to terminate a Management Agreement upon a sale of the Hotel is something an Owner must pursue with Manager. Manager is very invested in the Term of the Management Agreement and will have good and solid reasons to resist this termination right, at least not without a lockout period during which Owner could not sell the Hotel unless the Management Agreement was assigned to and assumed by the purchaser. This special right is intended to cover only a Sale of the Hotel to an unaffiliated third party. Some common stipulations include:

- prior written notice to Manager of Owner's intention to terminate the Management Agreement upon a Sale of the Hotel, with a significant period of transition time to reflect WARN Act considerations and any applicable collective bargaining agreement matters;
- a second notice to Manager of the actual date of termination of the Management Agreement upon the closing of a Sale of the Hotel, with a much shorter time between this notice and the closing, such as 15 days, and this notice of termination may be contingent upon the closing of the Sale of the Hotel;
- Owner's payment to Manager of all amounts due to Manager under the Management Agreement through the effective date of termination;
- Owner's payment to Manager of a pre-negotiated sale termination fee (the "Sale Termination Fee"), paid concurrently with the closing of the sale and as a condition precedent to the effectiveness of termination.

The Sale Termination Fee is often expressed as a sliding scale based upon fees received by Manager during the prior 12 to 24 months. Manager must also guard against finding itself managing a Hotel for an Owner that is a competitor of Manager. For this reason, if an Owner does not terminate the Management Agreement upon a Sale of the Hotel to a competitor, it may give rise to Manager's right to terminate the Management Agreement upon written notice to Owner, in which case Manager would be entitled to the applicable Sale Termination Fee.

The Sale Termination Fee is paid by Owner to Manager as compensation for the loss of the Management Agreement. To guard against the Management Agreement not being lost, the Management Agreement will provide that if Manager, or any Affiliate of Manager, accepts, at any time prior to an agreed upon date after the payment by Owner of the Sale Termination Fee, an engagement to manage a Hotel on the Site on behalf of the third-party purchaser (or any Affiliate of the third-party purchaser) that acquired the Hotel pursuant to the Sale of the Hotel that triggered the obligation to pay Manager the

Sale Termination Fee (but not any subsequent purchasers of the Hotel), within a short period of time after accepting that engagement, Manager must refund to Owner the entire amount of the Sale Termination Fee paid to Manager.

Special termination rights

In some special situations, Managers may attempt to negotiate into the Management Agreement the right to terminate the Management Agreement, upon prior notice to Owner, in the event of any suspension for a material period of time, such as 60 days, or any withdrawal or revocation, of any material governmental license or permit required for Manager's performance of its obligations under the Management Agreement or the operation of the Hotel in accordance with the terms of the Management Agreement that is not otherwise an Event of Default by Owner (in which case, Manager would be entitled to the rights and remedies provided to Manager in the Management Agreement upon the default of Owner) and such withdrawal or revocation is not reversed or otherwise cured within a reasonable period of time, such as 60 days, but only if the suspension, withdrawal or revocation is due to circumstances beyond Manager's reasonable control. A termination by Manager as permitted in these special situations does not entitle Manager to any termination fees due or payable to Manager under the Management Agreement.

In the marketplace today there are a number of new Brands available, and many Owners appreciate the potential value of a new Brand and a potential market differentiator with the potential to unlock profits. The same elements that offer potential from a new Brand also present challenges arising from a new Brand because the Brand is not as well recognized and perhaps untested. This leads some Owners to seek the right to terminate the Management Agreement if Manager abandons or fails to rigorously expand the Brand or the number of Hotels being operated under the Brand falls below a certain threshold. An Owner termination in this context does not lead to termination fees becoming due or payable to Manager.

Special provisions relating to interpretation and enforcement of the agreement

Management Agreements, particularly with the larger Brands, often include a mutual acknowledgment by both Owner and Manager that the long-term nature of the Management Agreement was a material factor in their respective determinations to commit to a long-term contract and make substantial contributions of time and money to the enterprise of Hotel management under the Management Agreement. This mutual acknowledgment also confirms that the parties have engaged in negotiations to avoid a wrongful termination of the Management Agreement, resulting in a Management Agreement that is terminable only in accordance with the express provisions of the Management Agreement itself. These types of provisions often have the following characteristics in the expression of the intentions of the parties:

- Owner acknowledges that Manager would suffer damage to its reputation if the Management Agreement were terminated by Owner in breach of the Management Agreement;
- Owner, as a material element of the bargain with Manager, waives any right or power otherwise available to Owner to terminate the Management Agreement other than in strict accordance with its terms;

- in any action or proceeding between the parties arising under or with respect to the Management Agreement, the Hotel or the relationship of the parties, each party unconditionally and irrevocably agrees to seek only its direct and actual damages and waives and releases any right, power or privilege either may have to claim or receive from the other party punitive, exemplary, statutory, or treble damages;
- both parties acknowledge that they are experienced in negotiating agreements like the Management Agreement, have been advised by legal counsel, and have specifically agreed that the express terms and conditions of the Management Agreement are intended to fully express and define the extent and limitations of the relationship between the parties;
- Manager will not be in default under the Management Agreement by reason of:
    - the Hotel's failure to achieve the financial performance expectations or income projections of Owner or the targets included in the Annual Budget;
    - the acts of Hotel employees, unless specifically addressed in the Management Agreement;
    - the institution of litigation or the entry of judgments against Owner with respect to Hotel operations;
    - any other acts or omissions that are not an event of default by Manager under the Management Agreement.
- Mutual waiver of a right to trial by jury of any actions under the Management Agreement that are not subject to expert resolution, mediation or arbitration;
- Prevailing party costs and expenses, including reasonable attorneys' fees and expenses, to the successful party in any adversarial proceeding.

Assignment

The long-term nature of most Management Agreements often does not fit well into the asset hold patterns of many Owners. Institutional Hotel Owners and private equity fund Hotel Owners often expect to sell the Hotel within a relatively short period of time, such as five to seven years, and real estate developers commonly prefer an even shorter hold period. This requires the parties to consider the circumstances under which the Hotel Owner can Transfer the Management Agreement, and if and when the prior consent of Manager will be required. (Remember to consult the *Glossary* with respect to the terms being explained here.)

Most Management Agreements provide latitude to Owner to Transfer Owner's interest in the Management Agreement without Manager's consent in the following circumstances:

- as collateral security for a Mortgage, and this will be in conjunction with an agreement to address subordination, non-disturbance, and attornment;
- to an Affiliate, so long as the Affiliate remains an Affiliate of Owner following the Transfer;
- to any person or entity acquiring by assignment Owner's interest in the Hotel if the assignee agrees in writing to be bound by the Management Agreement and assumes in writing all Owner's obligations under the Management Agreement from and after the effective date of the assignment.

Because Owner should have the financial capacity to provide working capital for the Hotel at times when Gross Revenues are inadequate to cover all Hotel Operating Expenses, an Owner will have certain limitations on Transfer, and may not Transfer the Management Agreement to any Person who cannot satisfy the following:

- has or can reasonably demonstrate a net worth threshold as established in the Management Agreement;
- is not known in the community as being of bad moral character, or is not in control of or controlled by Persons who have been convicted of felonies in any state or federal court;
- is a Competitor of Manager;
- is a Specially Designated National or Blocked Person.

Manager was engaged by Owner to manage the asset based on a selection process or criteria of Owner, so Manager too will find limitations on its right to Transfer its interest in the Management Agreement without the prior consent of Owner. Further limitations may also be imposed by the Mortgagee through the SNDA, because Mortgagee shall have also subjected Manager to underwriting and due diligence analysis in connection with the loan secured by the Hotel.

The Management Agreement's latitude for Manager Transfers without Owner's consent may include the following:

- so long as no Event of Default attributable to Manager has occurred and remains uncured, to an Affiliate, if the Affiliate remains an Affiliate of Manager following the Transfer;
- to any successor or assignee of Manager resulting from any merger, consolidation, transfer or reorganization, or to another Person which acquires all or substantially all of the assets of Manager or which, concurrently with the Transfer, becomes an Affiliate of Manager as long as the transferee expressly assumes by covenant in favor of Owner all the obligations of Manager under the Management Agreement.

What Manager cannot do without Owner's prior consent is the following:

- transfer the Management Agreement or a Controlling Interest in Manager to any Person who does not or is not, in Owner's reasonable judgment:
  - have sufficient operational expertise to fulfill Manager's obligations under the Management Agreement;
  - have substantially the same or better net worth than that of Manager and Manager's, if applicable when Manager is a SPE, parent company, immediately prior to such Transfer of the Management Agreement or a Controlling Interest in Manager;
  - have the right to utilize Manager's Trademarks and systems necessary to fulfill Manager's obligations under the Management Agreement, particularly if there is a Brand Manager;
  - have a positive reputation in the community as being of good moral character, and is not in control of or controlled by Persons who have been convicted of felonies in any state or federal court;

- a Specially Designated National or Blocked Person;
- a competitor.

In recognition of the number of Owners using a REIT structure, many Management Agreements provide that if Owner is or becomes a REIT, Owner will be able to Transfer the Management Agreement to a lessee of the Hotel, and Manager agrees to execute any amendment, amendment and restatement or replacement of the Management Agreement that Owner or lessee may require to incorporate the changes for compliance with Legal Requirements relating to the status of Owner as REIT.

Manager will also be permitted to undertake Transfers in the nature of collateral assignments of the right to receive fees and Reimbursable Expenses under the Management Agreement without the prior consent of Owner. In addition, any initial public offering or future transfers of ownership interests in Manager or its Affiliates on a publicly traded stock exchange will be excluded from Transfers prohibited under the Management Agreement.

Indemnification

Indemnification under a Management Agreement requires an analysis of two elements:

- What is the standard of the indemnification?
- What is the scope of the indemnification?

The standard for Manager's indemnification of Owner, which will include Owner (as stated in the Management Agreement) as well as its Affiliates and their respective agents, principals, shareholders, partners, members, officers, directors, and employees, will most often be limited to the uninsured fraud, willful misconduct, and gross negligence of Manager. That leads to the question of scope, and the most common construction in that regard is that Manager indemnifies Owner when the act or failure to act arises from Manager's corporate or home office personnel or the Key Personnel at the Hotel, meaning the General Manager, Chief Financial Officer, Director of Sales and Marketing or other senior executive position as identified (and commonly heavily negotiated) in the Management Agreement. The conduct of only a small and select group of people is imputed to Manager and deemed the conduct of Manager for purposes of Manager's indemnification of Owner.

With respect to Owner's indemnification of Manager, most Management Agreements express it as providing indemnification for anything and everything under the Management Agreement except as to specific acts or omissions for which Manager has agreed to indemnify Owner. This leaves Owner in the position of agreeing to indemnify Manager for any Losses occurring out of or by reason of the Management Agreement or otherwise arising in connection with the ownership, use, occupancy or operation of the Hotel. The ability to manage some of this potential exposure to liability for Owner through purchasing insurance is a key strategy that should not be overlooked.

Although not every Management Agreement contains further clarification as to the purpose and intent of the mutual indemnifications provided in a Management Agreement, it is helpful and wise to clarify that each party's indemnification of the other excludes any loss, cost, liability, expense, or claim actually covered by insurance. This may be insurance required to be maintained under the Management Agreement as well as any other insurance voluntarily placed by one of the parties. In addition, the indemnification provisions are intended to apply in connection with third-party claims brought against Owner or Manager by persons who are not a party to the Management Agreement and not claims of one party as against the other. Those types of claim that arise under the Management Agreement and the conduct of Owner or Manager to the other party should be adequately addressed within the Management Agreement itself under the default and default remedy sections.

All debts and liabilities arising in the ordinary course of business of the Hotel or otherwise in connection with the use, occupancy or operation of the Hotel are the obligation of Owner. Manager is not liable or otherwise responsible for any of Owner's debts or liabilities by reason of management of the Hotel, except for debts or liabilities arising because of Manager's fraud, gross negligence and willful misconduct.

The procedures for a party indemnified under the Management Agreement to act on the indemnification provided by the other party will be described within the Management Agreement. The basic elements of those provisions are as follows:

- prompt written notice by the indemnified party to the party required to provide indemnification;
- prompt assumption of the defense, by the party required to provide indemnification, including the employment of legal counsel selected by the party providing the defense;
- consultation with the indemnified party, which will have reasonable approval or consent rights as to the defense of the action;
- control of the defense and settlement of the action by the party providing the indemnification, including the right to negotiate settlement or consent to the entry of judgment; but, if any settlement or consent judgment contemplates any action or restraint on the part of the indemnified party, then any settlement or consent judgment requires the written reasonable consent of the indemnified party;
- the indemnified party's right, at its expense, to employ separate counsel and to participate in the defense of the action;
- the indemnified party's right to settle the action with respect to its own liability and with no requirement of the party providing indemnification to admit guilt or liability, but with the prior written reasonable consent of the party providing indemnification;
- if the party obligated to provide indemnification fails to use reasonable efforts to defend or compromise the action, the indemnified party may, at the expense of the party obligated to provide indemnification, assume the defense of the action and the party obligated to provide the defense must pay the charges and expenses of attorneys and other persons on a current basis within a short period of time after submission of invoices or bills therefor.

Consents and cooperation

During the course of a multiyear relationship between a Hotel Owner and its Manager there will arise innumerable situations that will require consultation and consent. One hallmark of a good situation between an Owner and Manager will be the atmosphere of collaboration and reasonableness. The Management Agreement can help foster this through a number of clear expressions of the desire and intention of the parties to act reasonably. For example, unless there is an express provision to the contrary within the Management Agreement, wherever the Management Agreement calls for the consent or approval of Owner or Manager, that consent or approval will not be unreasonably withheld, delayed, or conditioned, will be in writing and will be executed by a duly authorized officer or agent of the party granting the consent or approval. In furtherance of that standard for consent or approval, Owner will often agree to cooperate with Manager by executing leases, subleases, licenses, concessions, equipment leases, and service contracts pertaining to the Hotel. Of course, the Management Agreement will provide that those types of agreements must be negotiated in good faith by Manager and be permitted and approved by Owner under the Management Agreement, to the extent that Manager concludes, in Manager's reasonable judgment, that those agreements should be made in the name of Owner rather than the name of Manager.

Relationship

The relationship between the Hotel Owner and Manager is one of the most dynamic components of the Management Agreement and the business model whereby a third-party operates a Hotel for another. Each Management Agreement will be different and may depend on the nature of Manager as a large institutional Brand rather than as an independent third-party Operator. Some Management Agreements specify that in the performance of the Management Agreement, Manager acts as the agent of Owner. In other situations, the statement is that Manager is an independent contractor for Owner. Of course, when Owner and Manager are in a dispute, the nature of the relationship is often an element of that dispute, and the trier of fact, be it an arbitrator or a judge, will render his or her own opinion about what he or she considers the relationship to be regardless of what is stated in the Management Agreement.

Applicable law

The Management Agreement will be construed and governed by the laws of the jurisdiction set forth in the Agreement itself. One common approach is to apply the laws of the state in which the Hotel is located. However, there are some significant variations to explain. There is one state that has enacted legislation that addresses the Hotel Management Agreement and the concept of agency in Hotel Management Agreements. The State of Maryland enacted Title 23 of the Commercial Law Code to cover Hotels, meaning a Hotel or motel with more than 30 rooms for rent that is primarily used by transients who are lodged with or without meals, and the agreement for its management, defined as a written contract, agreement, instrument, or other document between at least two persons that relates to the management, operation, or Franchise of a Hotel. Under the laws of Maryland, if a conflict exists between the express terms and conditions of an operating agreement and the terms and conditions implied by the law governing the relationship between a principal and agent, the express terms and conditions of the operating agreement govern. This has led a number of Hotel management companies to establish or retain a legal nexus with the State of Maryland and specifically state in the Management Agreement express terms and conditions for the relationship of the parties and to modify what might otherwise occur under the historic common law principles of agency law. This technique may now be subject to further thought as a result of recent litigation under both Florida law and New York law ruling that Hotel Management Agreements are contracts for personal services and therefore subject to termination by Owner as the principal, but not without consequences such as liability of Owner to Manager for damages. For the purpose of understand the basic Hotel Management Agreement, it is important to not gloss over the selection of applicable law under the Management Agreement, as it can have meaningful consequences in the current market-place. Many Managers may wish to cling to Maryland in order to overrule common law principals of agency, while Owners may wish to designate New York or Florida in order to preserve the power to terminate a Management Agreement, even with the obligation to pay damages. (We have included the Maryland Statute at the companion website. Please visit www.routledge.com/cw/migdal to view the examples.)

Confidentiality

The complexity and detail of a Management Agreement lead the parties to consider the matters set forth in the Management Agreement to be confidential, and obligate the parties to ensure that the information is not disclosed to any outsiders without the prior written consent of the other party, subject to certain common exceptions as follows:

- as required by Legal Requirements;
- as reasonably necessary to obtain licenses, permits, and other public approvals;
- in connection with Owner's financing or Sale of the Hotel.

If an Owner is going to issue a prospectus, private placement memorandum, offering circular, or offering documentation related to the Hotel that includes the trademarks or intellectual property of Manager to interest potential investors in debt or equity securities, Manager's prior written approval will generally be required for use of the trademarks for that purpose. In addition, Managers will require that the prospectus clearly state that Manager has no responsibility for the prospectus. Owner will support this agreement to Manager with an agreement to indemnify, defend and hold Manager harmless from and against all loss, costs, liability, and damage (including reasonable attorneys' fees and expenses, and the reasonable cost of litigation) arising out of any prospectus.

Projections

Hotel Managers often provide Owners with operating projections of the Hotel. These projections are prepared well in advance of the development, conversion, or acquisition of the Hotel. Some projections are provided in advance of the execution of the Management Agreement. They are good faith estimates, but based on numerous assumptions that may or may not be true as the transaction progresses from Management Agreement execution to the opening date of the Hotel. To address this, Owner will acknowledge in the Management Agreement that any written or oral projections, proformas, or other similar information that have been provided by Manager to Owner prior to execution of the Management Agreement, or that will be provided during the Term, is for information purposes only. Manager does not guarantee that the Hotel will achieve the results set forth in the projections, and Owner acknowledges that the projections are based on assumptions and estimates, and unanticipated events may occur subsequent to the date of preparation of the projections. The reality that actual results achieved by the Hotel are likely to vary from the estimates contained in any projections is something that an Owner is often required to acknowledge within the Management Agreement.

Actions to be taken on termination

Despite the interest in having long-term management relationships, Hotel Management Agreements terminate. The Hotel Management Agreement should describe a procedure to effect the termination, such as the issuance of a written notice of termination, having an effective date as specified in the notice. The effective date is commonly not less than 60 days after the delivery of the notice to allow for the orderly transition of management of the Hotel, particularly with respect to employees. The Hotel Management Agreement should also make adequate provisions for what will transpire upon the termination of the Management Agreement and the organized and orchestrated transition from the current Manager to the successor Manager. Each of the following is often provided for in a Management Agreement, but caution should always be exercised to add, remove or modify provisions to accommodate the specifics of the situation:

- Within a reasonably short period of time after the Termination of the Management Agreement, often 30 days, Manager will prepare and deliver to Owner a "Final Accounting Statement," which is an accounting statement for the year to date, together with a statement of anything that Owner is obligated to pay to Manager under the Management Agreement to the date of Termination. Within 30 days after Owner receives the Final Accounting Statement, both Manager and Owner will make whatever cash adjustments are necessary pursuant to the statement. The cost of preparing the Final Accounting Statement is an Operating Expense, subject to an exception when the Termination occurs as a result of a Default by either party, in which case the defaulting party would pay the cost. The Final Accounting Statement may not truly be "final," so Manager and Owner will agree that there may be adjustments to occur later for matters that lacked available information when the Final Accounting Statement was prepared and that they will readjust the amounts and make the necessary cash adjustments when the information becomes available. This should have a sunset provision, so the Management Agreement will stipulate a date by which all accounts will be deemed final. This is a negotiated point, but between the first and second anniversary of the effective date of Termination is common. Of course there is often room for disputes, so the time period is subject to extension or tolling during the pendency of litigation or arbitration. If the Management Agreement provides for the resolution of certain disputes through expedited procedures before an Expert, any disputes concerning the correctness of the Final Accounting Statement should be referred to the Expert for resolution in accordance with the dispute resolution provisions of the Management Agreement.
- Manager must release and transfer to Owner, which might include an Owner designee, such as a purchaser if the Hotel is being sold, any of Owner's funds subject to Manager's possession or control with respect to the Hotel. This may be subject to an exception if funds are going to be held in escrow under the Management Agreement as part of the transition process.
- Manager must make available to Owner the Hotel's books and records, including those from prior years, that Owner will need to prepare its accounting statements for the Hotel for the year in which the Termination occurs and thereafter.

The Management Agreement should specify that the books and records will include employee records for Hotel Employees, other than records that must remain confidential under Legal Requirements or the regulations of any governmental authority or agency.

- Manager will assign to Owner or to the new Manager all operating licenses and permits for the Hotel that are in Manager's name, and the current licensee will cooperate in the transition of the licenses or permits to continue the operation of the Hotel under the licenses or permits. While this may vary by jurisdiction, the liquor and restaurant licenses must be examined and addressed. It is a very serious matter for a Hotel to be unable to serve alcoholic beverages during the transition from one Manager to another or one Owner to another. The adverse effect on the food and beverage and banquet business of the Hotel can be material, and the reputational impact can be even worse. Many, but not all, jurisdictions will either permit, or tacitly permit by ignoring, an agreement to permit a successor to operate under the existing liquor licenses for a limited period of time while the successor's application is pending. These Beverage Management Agreements or Interim License Agreements must be approached on a local level. Some jurisdictions have a favored form, and parties will always be better served by discovering and using that form. Some of the more critical and more common provisions of a Beverage Management Agreement include that:

  - the party using the license of another will provide insurance and indemnity protections during the interim period;
  - the party seeking the new license must apply and diligently prosecute the application;
  - only the licensee can purchase and serve alcohol under the license, so accommodations must be made for the use of employees and proper financial accounting;
  - the temporary use of the license will end on a negotiated date, regardless of whether the new license has been issued.

- If the Hotel Inventory and Fixed Asset Supplies includes items bearing the Trademarks of Manager, then Manager will usually have the option, and, occasionally, the obligation, to purchase those items at actual cost. The exercise of this option should have a time for its exercise, such as within 30 days after Termination. The Management Agreement should go on to provide for Owner's right to use those items until the supply is exhausted or destroy them, without liability to Manager or Owner of the marks.

- If the Hotel were designed and constructed with certain features that are proprietary to or commonly identified with Manager, even if protected by trademarks or service marks held by Manager, Owner will nevertheless be able to continue to operate the improvements on the Site without modifying the architectural design.

- All Manager's Proprietary Software and computer equipment utilized as part of Manager's centralized reservation system at the Hotel remain the exclusive property of Manager, who has the right to remove it from the Hotel without compensation to Owner. However, Manager's removal must occur in a manner that will minimize disruption to Hotel operations.

- Once a party receives notice of Termination, then continuing through the effective date of Termination, Manager cannot enter into any leases, licenses, concession agreements, maintenance contracts, service contracts, or any other form of agreement without the prior written consent of Owner. This should specifically exclude reservations or bookings of rooms, lodging, banquets, or other functions reserved in advance and made by Manager in the ordinary course of business, even though the event or booking is for a period after Termination, but commonly subject to Manager's obligation to collaborate and cooperate with Owner and any successor Manager concerning future event and group bookings.

- Upon a Termination of the Management Agreement, Manager will peacefully vacate and surrender the Hotel to Owner. It is often wise to have the Management Agreement anticipate that Owner might need to transition to new management, although this may be less available in a Management Agreement with a large Branded Manager. When an Owner can provide for it, the Management Agreement can include Manager's acknowledgment that upon a Termination, and before the effective date of Termination, Owner may enter into contractual arrangements for new management at the Hotel, and should Owner do so, Manager waives any claim against Owner or the third party for interference with contract. Manager should also agree to cooperate in the transition to a new Manager.

- Regardless of the reason for the Termination, Manager will produce a post-termination budget of all Manager's costs and expenses reasonably expected to be incurred by Manager. These costs and expenses are funded and paid by Owner in connection with terminating the Hotel, and commonly include costs for Employees whose employment will not continue following such termination, including severance pay, premiums related to health insurance continuation, accrued and unused paid time off (including, vacation, holidays, sick days, and personal days), unemployment compensation, employment relocation, multiemployer pension plan withdrawal liability under Title IV of ERISA, applicable taxes and/or penalties, exit fees from any employee benefit plan, and other employee liability costs arising out of the termination of employment of the Hotel Employees.

- Manager will cooperate with Owner or any successor Manager in connection with Owner's or the successor's hiring of Hotel Employees, and Manager will not hinder or frustrate the transition of employment of Hotel Employees.

- Manager will deliver to Owner all keys, passwords and other information necessary to facilitate a smooth transition of the operation of the Hotel to Owner or another Manager selected by Owner.

- Manager will deliver to Owner all financial data on the Proprietary Software and elsewhere relating to the operation of the Hotel in a format that can be readily uploaded at Owner's expense to a new software system selected by Owner.

- Manager will continue to diligently discharge all its obligations under the Management Agreement, including advertising and promoting the Hotel and actively seeking and accepting bookings notwithstanding that they are to occur after the expiration of the Term.

Trademarks and intellectual property

The use of intellectual property, trademarks and Branding is a common feature of the hospitality business today. This will be reflected in the Management Agreement as both Owner and Manager act to protect the integrity of their intellectual property and registered marks. While both Owner and Manager may possess intellectual property and Brand names or marks, it is most common for Manager to be the party that provides

Branding as an element of its management, and for that reason these components of most Management Agreements will be Manager driven. When an Owner engages a Manager, particularly a Branded management company, the Management Agreement is likely to be silent as to an Owner's marks, and address solely the marks of Manager. The language that follows will speak to Manager's marks, but the same can apply with equal effect in situations where Owner has its own intellectual property and marks to protect.

The Management Agreement provisions concerning intellectual property and marks will most commonly include the following concepts:

- Owner's acknowledgment to Manager that Manager is the sole and exclusive Owner of the defined Trademarks;
- the Trademarks will remain the exclusive property of Manager;
- all use of the Trademarks in connection with the Hotel, or as otherwise permitted under the Management Agreement, is for the benefit of Manager;
- nothing in the Management Agreement can be interpreted or construed to grant Owner any right of ownership in the Trademarks, other than use of the Hotel's name in certain reports to satisfy Owner's reporting obligations to its Mortgagee, investor or in the ordinary course of business;
- Owner cannot apply for registration of any Manager Trademark in any jurisdiction;
- Owner will not use the Trademarks without the prior written consent of Manager, and the standard for this consent will be Manager's sole and absolute discretion. Some illustrative examples of use that require consent include the following:

  - material prepared for the purpose of a Sale of the Hotel;
  - trade names adopted by Owner.

- Manager may continue to use its Intellectual Property in the ordinary course of its business or in any other manner, other than as may be restricted by the Management Agreement;
- Manager's Intellectual Property in Owner's possession is confidential and cannot be disclosed to any third party, without the prior written consent of Manager;
- when the Management Agreement terminates, all Manager's Intellectual Property will be removed from the Hotel by Manager, without compensation to Owner;
- when the Management Agreement terminates, Owner will have additional obligations to do the following:

  - immediately place coverings over or render not visible to the public any signs containing the Trademarks;
    - remove any signs or similar identification from the Hotel within a very short period of time, often no later than 30 days after the date of Termination;
  - immediately either remove from the Hotel all Fixed Asset Supplies, Inventories, and other items bearing the Trademark, remove all Trademarks from such items, or use the Inventories, Fixed Asset Supplies, and other items bearing the Trademark exclusively in connection with the Hotel until they are consumed.

- the Management Agreement will empower Manager to take certain actions should Owner fail to comply with the Management Agreement's provisions concerning Trademarks. For example, if Owner fails to remove signs bearing the Trademarks, Manager will have the right to do so at Owner's expense. It is often important to be careful with Trademarks and all elements of Manager's intellectual property because it is possible that the laws of a different jurisdiction will govern the Trademarks and other components of intellectual Property.

Remember that the Management Agreement can also include Intellectual Property of Owner. In those situations much of what is addressed here will be repeated in the Management Agreement for the benefit of Owner. In addition, it may be appropriate to also cover these points:

- Manager's ability to use Owner's Intellectual Property for Manager's national marketing programs in accordance with the Marketing Plan that Owner and Manager have devised for the Hotel;
- Manager's covenant to take all reasonable steps to ensure that Owner's Intellectual Property remains confidential and is not disclosed to anyone other than Manager's employees at the Hotel;
- Upon Termination of the Management Agreement, Manager will make available to Owner the Hotel Financial/Guest Data that are stored in Manager's Intellectual Property, subject to Manager's ability to retain a copy of the Hotel Financial/Guest Data following Termination.

Because the provisions for the protection of Intellectual Property are so crucial and not capable of being adequately protected through the payment of money damages on a breach, the Management Agreement should provide each of Owner and Manager the right to injunctive relief in addition to any other right or remedy available at law or in equity to enforce those Management Agreement obligations, including all costs of enforcement and reasonable attorneys' fees.

Because so much attention is paid to the food and beverage components of a Hotel, and there is the potential for significant financial gains or financial losses if the food and beverage component is not properly incorporated into the Hotel, the guest experience, and overall operations, care should be exercised in addressing special situations, such as when a food and beverage concept or venue has been designed and created just for a particular hotel. The Management Agreement or a related ancillary agreement should address the intellectual property associated with that venue, as well as any developed amenity, to clearly describe the party that owns and holds the intellectual property, be it Manager, Owner or as a jointly held asset, and how the parties would handle the termination of the Management Agreement or the expansion of the intellectual property, should the enterprise prove successful, or the termination of the enterprise should it prove not successful or should a party simply desire to exit the enterprise.

Hotel guest records

The treatment of the Hotel's Guest Records in the Management Agreement can take many forms and defies a single definitive explanation. That being said, there are certain common elements in the understanding of this aspect of Owner and Manager or Franchisor and Franchisee relationship that can be instructive. Most often, the manner of treatment will depend on whether the Hotel is affiliated with a Brand, through management or franchise. This, perhaps obvious, point cannot be overstated. One of the assets that Brands bring to the relationship is their brand strength, reach, and power, as executed through the use of the Brand's intellectual property, proprietary information, registered logos and marks, loyalty and rewards programs, and guest information and data. None of this will be offered to an Owner or Franchisee when the Brand departs the Hotel. That being said, an Owner or Franchisee still needs to operate the Hotel upon the departure of the Brand, and should negotiate for the guest records necessary for a transition from one Manager to another with continuous operation of the Hotel.

An analysis of the defined terms in the Management Agreement is the first step to understanding the respective rights and obligations of Owner and Manager as to Hotel Guest Records. Any number of defined terms may be relevant to this analysis. For example, one sample formulation of "Hotel Guest Records" might mean guest records, profiles, histories, contact information, and preferences gathered by the Hotel based on the guest's stay or information provided by the guest during such stay at the Hotel (whether maintained as a separate database or as part of an integrated database utilized by Manager, the parent company of the Brand or any of their respective Affiliates). This Hotel Guest Data, gathered from the guest's stay at the Hotel, can generally be jointly owned by Manager and Owner. Care must be exercised to distinguish information gleaned during a guest stay from information that Manager possesses based on a loyalty or reward program or from a stay by the guest at another hotel within Manager's System or hotels. This type of information not derived from the Hotel will be proprietary to Manager and not available to Owner upon the expiration or termination of the relationship with Manager.

Generally, Manager, and its affiliates, have the right to use the Hotel Guest Records in any manner, subject, of course, to all Legal Requirements, which is intended to include the current state of the law as to the protection of consumer information. Owner's use of the Hotel Guest Records is often restricted both during the Term and after the Term to use in connection with the Hotel, also subject to Owner's compliance with all Legal Requirements. Sometimes, Manager will grant Owner a license to use the Hotel Guest Data after the Term. A common way to express this is that, anything to the contrary in the Management Agreement notwithstanding, upon termination of the Management Agreement for any reason, Manager will grant to Owner the right and license to copy and utilize for Hotel purposes and consistent with applicable laws and regulations all Hotel Guest Records created during the Term and relating to guests' stays at the Hotel. The License survives the expiration or termination of the Management Agreement, but is limited to the operation of the Hotel. Whether Owner owns the Guest Data jointly with Manager or obtains a perpetual license to use the Guest Data, it is generally only in connection with the Hotel. Owner is not permitted to otherwise monetize the Guest Data.

Expert decisions

Owners and Managers often segregate specific matters that may be determined by an Expert rather than through mediation, arbitration or litigation. The use of a single Expert to promptly resolve certain specific matters can allow the parties to break an impasse through a proceeding that is intended to be expeditious, have a limited scope, and employ a single arbitrator for prompt resolution. There is a significant range of variation in this area that is often based upon the underlying policy or predilections of one of the parties. Adverse experiences in one form of dispute resolution or the other can direct policy for the future based entirely on that single negative experience. Nevertheless, many Management Agreements will require that certain matters be referred to an Expert for determination, such as certain Budget disputes and revisions to the Competitive Set. Some Management Agreements may say that all matters will be referred to the Expert, but that is not the majority view.

Expert determination provisions often include the following:

- the use of the Expert will be the exclusive remedy of the parties for either the entire Management Agreement, although this is not the most common approach, or selected issues under the Management Agreement, such as relating to the Budget, and neither party is permitted to attempt to adjudicate those disputes in any other forum;
- the decision of the Expert is final and binding on the parties and neither party will challenge the determination of the Expert;
- each party can make written submissions to the Expert, and the other party can then make comments;
- the Expert will have access to all books and records relating to the issue in dispute with the cooperation of the parties;
- the Expert will have the authority to direct who will pay the costs of the Expert and the proceedings, unless otherwise provided for in the Management Agreement;
- the Expert may direct that the costs be treated as Operating Expenses.

In accepting the engagement to serve as the Expert, the Management Agreement may specify certain obligations of the Expert such as the following:

- to notify the parties in writing of the decision within a stipulated period of time from the date of the selection of the Expert;
- to establish a timetable for the making of submissions and replies.

Area of Protection

An Area of Protection or Exclusive Territory limits the ability of Manager to open another Hotel under the same Brand name during a negotiated period of time and in a negotiated geographic area. This is an Owner's responsibility to require and negotiate. Each component, duration, geographic scope and definition of a competing Hotel is likely to be discussed at length by the parties. It is unlikely that a large Branded management company will offer exclusivity beyond the specific Brand that is the subject of the Management Agreement, but on a case-by-case basis, that issue may be worthy of discussion. Some exclusive territory during the entire Term is the ideal from Owner's perspective. Creativity to fit the location of the Hotel and the geographic area is important. Some Management Agreements include multiple Exclusive Territories that may collapse after a term of years, leaving a smaller territory in place.

An Exclusive Territory will not prohibit Manager from participating in other Hotels bearing any other Brands of Manager. In addition, if Manager engages in a merger, consolidation or acquisition of a Brand of Hotels consisting of more than a negotiated number of Hotels, intended to cover a portfolio acquisition by Manager, then the Hotels that are part of that transaction will be excluded from the Exclusive Territory of the Management Agreement.

Quiet enjoyment

Although the Management Agreement does not grant Manager an interest in the real estate, meaning the Hotel, Manager will often negotiate into the Management Agreement an Owner's covenant of quiet enjoyment in favor of Manager. Manager will be able to, and should be able to, peaceably and quietly possess, manage and operate the Hotel throughout the Term. That is part of what a Manager negotiates for and what an Owner desires since Owner does not want to operate the Hotel and desires to engage an operator to act on behalf of Owner. Owner covenants to undertake, at its own expense, to prosecute any appropriate action, judicial or otherwise, to assure Manager's peaceful and quiet possession and enjoyment. This relatively rational and often innocuous provision in a Hotel Management Agreement raises some very interesting issues. The concept of quiet enjoyment is essentially a real estate concept. In the context of a Hotel Management Agreement, there should be no interest in favor of Manager in Owner's real estate, so the use of real estate-based concepts and language may be met with resistance from Owner. The better approach to permit both Manager and Owner to receive the benefit of their bargain is for the Management Agreement to express that Manager is granted by Owner the right to exclusive possession and control of the Hotel during the Term to manage the Hotel in accordance with the Management Agreement, but that Manager does not possess, and disclaims and waives, any real estate interest in and to the Hotel.

No interest in real estate; recordation

As described earlier, Owner's covenant of quiet enjoyment to Manager is best described in terms of contract and not in the granting of an interest in the real estate itself. This is balanced by Manager's acknowledgment that the Management Agreement does not constitute or grant any interest in the real estate constituting the Site and the Hotel. Whether a memorandum of the Management Agreement will be recorded in the real estate or other public records applicable to the Hotel is a point to be negotiated, and it is common for Owners to resist Manager's desire to record a Memorandum of Management Agreement. Many large Branded management companies insist that a Memorandum of Management Agreement be recorded. The recording of a memorandum among the applicable land records is viewed as important to the Branded Manager, but expect to see Owner resistance referencing Manager's lack of an interest in the real estate as support for the argument.

4 Essential Management Agreement Case Law

Hotel Management Agreements are complex instruments that involve the interplay of a diverse set of assets and liabilities, and are frequently executed for extended periods of time. When a dispute arises under such circumstances, the stakes are high whether considered in the context of a negotiated agreement between the parties or an adversarial proceeding when all else fails. When Manager and Owner are unable to resolve the issues between them and litigation (or arbitration) ensues, the result often turns on the interpretation of the relationship between those parties, and what forms of relief are available to the aggrieved party as a result of the claimed breaches of contract by the other. New York, Florida, and California have produced the leading cases in this area on these issues. This chapter will examine *Woolley v. Embassy Suites* and its progeny, and their effect on the resolution of disputes between Managers and Owners, as well as some of the remedies available in these types of disputes.

Termination and relevant contractual language

In the cases discussed in this chapter, a typical dispute arises from an Owner's desire to terminate Manager of its hotel (whether for cause or not). Hotel Management Agreements contain strict termination clauses that are designed to protect a Manager's rights in the event that an Owner pursues a claim for breach of contract. For example, the typical Hotel Management Agreement provides that when Owner believes Manager has committed a material default, Owner must provide Manager with a written notice of default that also carries an opportunity for Manager to effect a cure of the alleged default, and the time to effect that cure may be subject to extensions based on the circumstances of the default. Only after Manager fails to cure the default may Owner seek to enforce its rights and remedies under the Management Agreement, which almost universally will include termination of the Management Agreement. In most situations, Owner then must provide Manager with a further notice that Owner is seeking termination of the Management Agreement. Many Management Agreements further limit the right to seek termination by providing that the Management Agreement is not subject to termination if there is pending a bona fide dispute regarding a material default that has been submitted to arbitration, when the Management Agreement calls for the arbitration of certain disputes, which many Management Agreements indeed mandate. These requirements may produce extended disputes while an Owner may be seeking an expedited resolution. In the past, some Owners have either gone through the process of obtaining a preliminary injunction to remove Manager from their property or have resorted to extrajudicial means to remove Manager. Extrajudicial actions have the potential to impact a Hotel Management Company's fundamental interests by resulting in lost revenue streams, negative reputation effects, and upheaval to an overall market strategy and business plan.

To protect these interests, Hotel Management Companies began to bargain for certain provisions in Management Agreements to afford them extra protections and buffers before the remedy of termination could be effected by an Owner. These provisions specifically relate to the availability of remedies and the characterization of the relationship between Owner and Manager. A fair reading of these provisions is that the effect is to impede termination of Manager and grant Manager the right to obtain specific performance and injunctive relief. The Model Hotel Management Agreement purports to afford this extra protection through the following provisions:

- An explicit acknowledgment that Manager would not have entered into the agreement or made significant monetary and temporal investments necessary to perform the agreement unless the term was for a substantial period of time, and therefore, the limitations on remedies and rights in termination had been specifically and extensively negotiated.
- An agreement that Manager's damages would be impossible to determine in the event of a breach and termination. This provision also sets forth liquidated damages in the event of a breach and a methodology for computing actual damages because the damages are difficult to ascertain.
- Most importantly, an acknowledgment by Owner that Manager would suffer damage to its brand and reputation, that monetary damages would not be an adequate remedy, that Manager's interest in the Hotel is that of an agency coupled with an interest, that Owner waives any right of termination outside such rights set forth in the agreement, and that Manager has the right to seek specific performance of the agreement and injunctive relief without bond. This same provision further warrants that each party is a sophisticated entity with experience in negotiating similar agreements, and have received counsel regarding these provisions.
- A separate provision providing that where there is a conflict between the terms and conditions of the agreement and implied agency law, the agreement's terms shall govern the dispute. It then reiterates the availability of specific performance as a remedy, regardless of any agency relationship between Manager and Owner.

These provisions can be found in Article 16 of the Model Hotel Management Agreement, and, typically, the Management Company requires Owner of the hotel explicitly to initial them in the Agreement. Many of the clauses enumerated above were developed specifically to respond to the various holdings of *Woolley v. Embassy Suites* and the cases that rely on it.

Injunctive relief

When resolving disputes that arise from alleged breaches of a Hotel Management Agreement, courts are often confronted with requests for injunctions that may alternatively order a Manager to vacate a Hotel, prevent an Owner from removing a Manager from a Hotel, or order an Owner to allow a Manager to reenter a Hotel. Managers sometimes seek injunctions to maintain the validity of their Management Agreement to protect their extended terms and interests in managing a particular piece of hotel real estate.

The courts have used traditional injunctive principles to resolve these issues. To grant an injunction, a court must find that:

- The party seeking the injunction is likely to prevail on the merits.
- The party seeking the injunction would suffer irreparable harm if the injunction is denied, and damages would not be sufficient to make the aggrieved party whole.
- The balance of hardships between the party seeking the injunction and the party against whom the injunction is sought leans towards the party seeking the injunction. In other words, that a denial of the injunction would damage the party seeking the injunction more than granting the injunction would damage the party against whom the injunction is sought.
- That granting the injunction does not harm the public interest.

With very limited exceptions, Managers have had no success in seeking prohibitive injunctions preventing Owners from removing them from Hotels after the termination of the Hotel Management Agreement, nor have they succeeded at obtaining mandatory injunctions allowing them to reenter a hotel. In addressing the element of irreparable harm, Managers have asserted that they would suffer irreparable harm through the misuse of their intellectual property and confidential and proprietary property, loss of goodwill, damage to reputation and brand name, and overall effects in the hospitality industry of permitting the unilateral termination of management companies. The courts have found these arguments unpersuasive because Manager's primary interest in the Hotel Management Agreement is its management fee, which is calculable. On the other hand, Owners have succeeded in obtaining injunctions removing Managers from their Hotels based on theories of continuous trespass and loss of control of reputation and goodwill.

Agency and personal services contracts

In determining whether a Manager would succeed on the merits, several courts have narrowed the question to whether a particular Hotel Management Agreement has created an agency coupled with an interest. Moreover, in order for injunctive relief to be available, an agreement must not be a contract for personal services, as such contracts historically have not been enforceable by specific performance.

The courts have relied on the principles of agency set forth in the Restatement of the Law of Agency in determining whether a Hotel Management Agreement creates an agency coupled with an interest. The Restatement states that an agency is a fiduciary relationship between a principal and an agent where the agent shall act on behalf of the principal and is subject to the control of the principal. The control prong of the agency relationship has proven problematic because Hotel Management Agreements often grant Manager wide discretion in the operation of the Hotel while significantly limiting an Owner's approval right over certain actions. It also provides that an agency relationship is not dependent on how the parties characterize their relationship, or on the form of the agreement, but on the substance of the relationship and whether it meets the basic agency requirements. Moreover, and this is the essential component of the agency relationship, the principal always retains the power to terminate the agency, unless it is coupled with an interest.

An "agency coupled with an interest," is also referred to, and can be remembered as, a "power given as security." It is an agency power that is given to the agent to protect a vested interest possessed by the agent in the subject matter of the agency. In the context of the hospitality industry, it means the Hotel itself. Many Hotel Managers have aggressively pursued the agency coupled with an interest characterization because it renders the agency irrevocable, and possibly makes available to Manager other remedies, such as injunctive relief. Agencies coupled with interests are traditionally created for the benefit of the agent and not the principal and are used to secure a performance of a duty or property interest, such as a creditor's interest in a property put up for sale.

In contrast, Owners have advocated for characterizing Hotel Management Agreements as personal service agreements. Personal service agreements are uniformly unenforceable by specific performance. Personal services are services that are non-delegable, have a certain degree of trust between the parties, and are difficult to judge with regards to quality of the performance.

Woolley v. Embassy Suites remains one of the benchmark cases for the hospitality industry, and sets out the fundamental holding that a typical Hotel Management Agreement does not create an agency coupled with an interest, and even if that could be established, the agreement would not be enforceable by specific performance because it is a personal services contract. The facts and circumstances of *Woolley* left many open questions regarding what might constitute a sufficient interest to create an agency coupled with an interest, but, even so, the cases that followed *Woolley* also denied relief to Manager despite factual distinctions between them and *Woolley*. The courts have rejected assertions that intellectual property interests, ownership interests in the Hotel held by the Brand's affiliates, the prevention of "havoc in the industry," loss of position in the industry, lease interests, purchase options and rights to quiet enjoyment are sufficient to establish an agency coupled with an interest, because they are neither present nor vested property interests in the Hotel itself, the subject matter of the agency.

Most recently, Florida and New York courts have held that Hotel Management Agreements are not agencies, but are personal services contracts, and therefore injunctive relief is denied entirely to a Manager seeking to prevent the termination of its Management Agreement. These courts emphasized that the owner ceded significant control to the Managers. Therefore, objective evaluation of the Hotel Management Agreements was difficult, a defining aspect of personal services contracts.

The landmark cases

Woolley v. Embassy Suites, Inc., 227 Cal. App. 3d, 1520 (Cal. Ct. App. 1991)

Woolley v. Embassy Suites was a California Appellate Court case that has become widely cited in other cases involving disputes over hotel management agreements. It addressed the availability of injunctive relief, agency law, and personal services as they relate to the specific performance of hotel management agreements. While the facts of the case limited the reach of its holding, the principles set forth in *Woolley v. Embassy Suites* have become foundational.

The owners were the general partners of 22 partnerships that owned 22 hotels. Embassy Suites was the franchisor for each hotel and managed 17 of the hotels as their operator. The owners sued Embassy Suites, asserting that it had materially violated the terms of their hotel management agreement by making expenditures in excess of budget allocations, and claimed breach of contract, negligence, fraud, and other wrongdoing. The hotel management agreement did not purport to limit termination rights and explicitly declared that it established an agency relationship between owner and manager. It also granted the owner approval rights over the hotel's budget, the retention of legal counsel, and hiring long-term employees, and also required the manager to submit monthly financial reports to owner. Finally, the hotel management agreement mandated that all disputes be submitted to arbitration for resolution.

After the owners issued a notice of termination, Embassy Suites applied for an injunction to restrain the owners from terminating the hotel management agreement, and the trial court granted that relief stating that Embassy Suites would suffer irreparable harm in the event that the agreement was terminated prior to the conclusion of the arbitration because the dispute would be rendered moot by the termination. The owners appealed the grant of the injunction.

The *Woolley v. Embassy Suites* court confronted four issues in the context of whether granting a preliminary injunction was appropriate:

1   whether the California Civil Code modified the substantive requirements of provisional relief by allowing injunctive relief in aid of arbitration;
2   whether the hotel management agreement established an agency between the owners and Embassy Suites;
3   whether that agency was irrevocable as an agency coupled with an interest; and
4   whether the hotel management agreements were personal services contracts unenforceable by specific performance, therefore foreclosing similar provisional relief such as an injunction preventing the owner's termination of the agreement.

The court analyzed the dispute using the traditional principles governing the grant of a preliminary injunction, namely, whether the aggrieved party had a probability of success on the merits, whether there was an irreparable injury, and whether damages would be an inadequate remedy.

The Hotel Management Agreement established an
agency between Manager and Owner

To determine whether Embassy Suites' contract was revocable under agency principles, despite its terms, the court had to determine whether an agency relationship existed. In contrast to later cases in which the manager conceded the agency relationship, Embassy Suites argued that despite the hotel management agreement's explicit terms that the agreement created an agency, an agency relationship did not exist because the owner entrusted the entire operation of their hotels to Embassy Suites, undermining the control aspect of agency relationships. Embassy Suites sought this characterization to preserve the right to pursue arbitration, otherwise the agreement would be revocable at the will of the principal, as are most agencies. The court dispensed of these arguments, relying on the hotel management agreement's explicit language, and the significant approval rights granted to the owner. Moreover, the court noted that for control to exist, it did not need to be exercised, but the *mere existence* of the right to control the agent would suffice.

The agency was revocable

Regarding the argument that Embassy had an agency coupled with an interest, the court held that Embassy's financial interests and the long-term economic interest in enhancing Embassy Suites' reputation were insufficient to create an agency coupled with an interest. In order for an agency to be coupled with an interest, the court held that it must be created for the benefit of the agent to protect a title or right in the hotel or to secure the owner's performance to the manager, independent of the agency relationship (as opposed to the subject matter of the agency, the hotel). Moreover, the agent must have a specific, present, and coexisting beneficial interest in the subject matter of the agency. Monetary consideration, or here the hotel management fees, is not sufficient to establish an interest. The court also dispensed of the manager's argument that it had an interest in enhancement of its reputation by interpreting the related franchise agreements and hotel management agreements. The manager's name was transferred through separate franchise agreements that were all severable from the hotel management agreement, further evidenced by the fact that some of the owners' "Embassy Suites" hotels were managed by other companies. Accordingly, the manager's interest in seeing its brand grow was not the type of property interest necessary to create an agency coupled with an interest.

Injunctive relief

The *Woolley v. Embassy Suites* court further elaborated that a preliminary injunction was improper because the manager's harm was not irreparable, an adequate remedy at law existed, and the agreement was a personal services contract that was unenforceable by injunction. The damages were readily calculable because the contract set out the fee percentages to which Embassy Suites was entitled. While the court recognized loss of reputation as a possible claim, it held that it could be ascertained through expert testimony. Finally, the grant of a preliminary injunction was improper because specific performance would not have been available in a determination on the merits because the hotel management agreement was a contract for personal services. Because the hotel management agreements required the exercise of special skill and judgment in the administration of a complex enterprise, a certain degree of trust between the parties given the degree of discretion involved in the operation, and the difficulty of judging such performance, the court ruled that the hotel management agreement was a personal services contract. The court specifically noted two policy considerations in refusing specific performance in this case, judicial economy and the 13th Amendment's bar against involuntary servitude:

> There are a variety of reasons why courts are loath to order specific performance of personal services contracts. Such an order would impose upon the court the prodigious if not impossible task of passing judgment on the quality of performance. It would also run contrary to the Thirteenth Amendment's prohibition against involuntary servitude. Courts wish to avoid the friction and social costs which result when the parties are reunited in a relationship that has already failed, especially where the services involve mutual confidence and the exercise of discretionary authority. Finally, it is impractical to require judicial oversight of a contract which calls for special knowledge, skill, or ability.

The injunction was overturned, and the owners were permitted to proceed with their termination of Embassy Suites. However, the circumstances of *Woolley v. Embassy Suites'* holding, particularly as it related to whether an agreement created an agency coupled with an interest, limited its direct applicability to other cases. Unlike the hotel management agreement in *Woolley v. Embassy Suites*, many management agreements today explicitly provide that they create an agency, and in particular, one coupled with an interest. Later litigants have asserted a wide range of interests, ranging from an affiliate's ownership interest in the hotel to expenditures made for capital improvements. Despite these interests, courts have continued to reject that a hotel management agreement creates an agency coupled with an interest.

Agency coupled with an interest

Because in *Woolley v. Embassy Suites* the hotel management agreements in question only created interests in "enhancing brand reputation" and the management fee, it left open the question of whether hotel management agreements were categorically barred from creating agencies coupled with interests. The following cases, *Pacific Landmark Hotel Ltd. v. Marriott Hotels Inc. (Pacific Landmark)*, *Government Guarantee Fund of Republic of Finland v. Hyatt Corp (Skopbank v. Hyatt)*, and *Fairmont v. Turnberry*, expanded on what constitutes an agency coupled with an interest in the context of the termination of hotel management agreements. Like their predecessor, their analyses were framed around whether to grant or deny an injunction. Moreover, the underlying hotel management agreements were much different than the hotel management agreement that controlled the outcome in *Woolley v. Embassy Suites*.

Pacific Landmark

In *Pacific Landmark*, the California Court of Appeals confronted a much more involved relationship between the owner and manager. The owner of a California resort alleged that Marriott had breached the hotel management agreement, and sought to terminate Marriott as the hotel's manager. The hotel management agreement provided that the relationship between the owner and manager established an agency coupled with an interest, as does the model document at the companion website. (Please visit www.routledge.com/cw/migdal to view the document.) It further provided that the agreement did not create a partnership, lease, or joint venture. Besides the explicit contract provision, which alone cannot create an agency coupled with an interest, one of Marriott's affiliates had secured a 5% ownership interest in the resort in exchange for lending the owner 15 million dollars and making eight million dollars of capital contributions in the hotel. In addition to the ownership interest, the affiliate also secured title over the tax benefits generated by each portion of the resort, and secured the loans by deeds of trust in the property. The trial court granted an injunction restraining the owner from terminating Marriott, holding that this relationship had created an irrevocable agency.

The appellate court followed the formulation set forth in *Woolley v. Embassy Suites* and principles of legal separateness of entities in determining whether the hotel management agreement created an agency coupled with an interest. In addition to the traditional requirement that the agent must have a specific, present, and coexisting beneficial interest in the hotel, the *Pacific Landmark* court held that when the power and interest are transferred at the same time, they must be coupled in time, in subject matter, in the same person, and in source. The court ruled that the interests that Marriott's affiliate possessed in the hotel were irrelevant, stating that Marriott could not avail itself of alter-ego theories to disregard the separate incorporation of its own entities. In other words, it was not permitted to pierce the corporate veil of its own subsidiaries because such measures are only taken when the corporate form is abused to harm the rights of third persons. The court also noted that such forms are used to insulate corporate parents such as Marriott from liability.

Marriott also claimed that the contracts executed in connection with the hotel management agreement gave rise to an aggregate overall interest. However, the court ruled that these aggregate interests were each individually insufficient to establish an interest in the hotel. Rather than being explicitly interdependent, the agreements each contained an entire agreement clause that isolated them from one other. Accordingly, none of the interests asserted by Marriott had unity in the same person, time, or subject matter. The court also noted the distinction between the power to revoke the agreement and the right to revoke the agreement. While an agency might be irrevocable by its terms, and therefore strips the owner of the contractual right to revoke the agency, the power to revoke the agency continues. In the *Pacific Landmark* case, the California legislature codified the power to revoke an agency in the California Civil Procedure Code. However, as other cases have noted, this independent power may also be found in the common law of agency. When an entity exercises a power of revocation where it does not have the right, it may give rise to damages resulting from the termination, but the power to effect the termination of the agency remains.

Because of *Pacific Landmark*'s specific facts, namely that the interests asserted were secured by a third entity with separate legal personhood from the manager, the courts had further occasion to elaborate on its principles. In later cases, the interests asserted would be held by the same entity that executed the hotel management agreement, eliminating that barrier to obtaining ultimate injunctive relief. While later interests did not include a direct ownership interest, they would implicate forms of capital investments, major contributions of a management company's resources to the success of the hotel as an overall business operation, and SNDAs. The language in hotel management agreements also continued to evolve, creating specific language protecting a manager's rights during the course of termination and explicitly setting forth the manager's expertise and resources as consideration for extended terms.

Government Guarantee Fund of Republic of Finland v. Hyatt Corp (Skopbank v. Hyatt)

In *Skopbank v. Hyatt*, the Third Circuit relied on both *Pacific Landmark* and *Woolley* in holding that the manager's stake in the overall operation of the business of the hotel, as evidenced by capital contributions, did not constitute a sufficient interest in the hotel to create an agency coupled with an interest.

The facts of *Skopbank v. Hyatt* illustrate the risks involved for a manager taking over a hotel. Hotel management companies are structured to limit their exposure to liability through operating, rather than owning, hotels. However, the operation of such hotels requires the significant expenditure of resources, including putting at risk the hotel management company's intellectual property, capital, and, most importantly, their brand name. Because hotel management companies rely on their brand name to generate their business, they may be reluctant to lend it to an enterprise lacking a positive operating history in the absence of substantial financial and economic incentives. Accordingly, when taking over a troubled resort, they may exact specific concessions regarding compensation and the operation of the hotel during negotiations to protect their interests.

Hyatt executed management agreements with the original owner, and the owner's lender, Skopbank. Skopbank lent the owner a significant amount of money, and the owner subsequently sought Hyatt's management expertise to make the enterprise profitable. Because of the resort's prior performance, Hyatt was concerned about committing its name. To justify the risk, Hyatt extracted contractual concessions giving Hyatt significant management control and profit sharing. The parties mutually agreed that Hyatt's reduction in front-loaded management fees constituted a capital investment, and that it had specifically bargained for a thirty year term and otherwise would not have entered into the agreement. The management agreement also provided that it was not subject to forfeiture or termination except within its express terms. Finally, Hyatt extracted a non-disturbance covenant from Skopbank, warranting that a change in ownership would not affect the underlying management agreement.

Skopbank foreclosed on the property because the owner failed to make payments on its debt, and a new owner, 35 Acres, attempted to terminate the management agreement. Despite 35 Acres's attempts to remove Hyatt from the property, Hyatt remained.

While the court decided the question of whether a hotel management agreement created an agency coupled with an interest under the law of the Virgin Islands, it noted that the result would likely be the same under Ohio or New York Law.

The *Skopbank v. Hyatt* Court held that the key inquiry was whether the agent had a proprietary interest in the subject matter of the agency relationship itself. Hyatt attempted to analogize its commitments to other cases where agents had made substantial contributions or investments in the business. Here, the asserted interests in the operation of the business were foreclosed by the language of the hotel management agreement itself, which specifically stated that it did not create a joint venture or partnership, or any other relationship. For the same reason, the court rejected that the related use of trademarks, service marks, use of proprietary and confidential information, systems for sales and marketing, and the integration of the hotel into the worldwide Hyatt business structure constituted sufficient interests. Rather, these were incidental to the performance of the agency, and, as such, they did not make it irrevocable. The court also ruled that it never was a true partner because it never bore the risk of loss. Its "capital expenditures" were merely incidental to the sale of management services.

Hyatt also attempted to specifically distinguish itself from *Woolley v. Embassy Suites* and *Pacific Landmark*. It alleged that because Hyatt had permitted the use of its intellectual property in connection with the hotel "*only because it insisted upon and received agency powers to manage the resort.*"

Hyatt attempted to claim that there was a unity of entities, in contrast to *Pacific Landmark*, and therefore an agency coupled with an interest had been created. However, the court rejected this claim for the same reason, that the management agreement by its terms did not create a joint venture or partnership. The court rejected the brand and reputation claims for the same reasons present in *Woolley v. Embassy Suites*. Finally, the SDNA was held to be an insufficient interest, since, although it may have evidenced the parties' intent that the management agreement be irrevocable (and Hyatt's foresight), it did not reach the substance of the question. Despite its careful contractual language, Hyatt found itself in the exact position it desired to avoid because the management agreement could not be found to substantively create an agency coupled with an interest.

Finally, the Third Circuit raised the question of personal services contracts and the unavailability of injunctive relief for breach of such contracts. Relying on *Woolley*, it stated that "[t]he plain language of the Management Agreement shows ... that the agency was created for the benefit of the owner, not Hyatt, and that Hyatt's sole interest in the Management Agreement is its right to compensation. As such, the Management Agreement was a personal services contract which cannot be specifically enforced." The specter of the personal services contract would be raised again with particular force later in the *Eden Roc* and *RC/PB v. Ritz-Carlton* cases, discussed later in this chapter.

FHR TB LLC v. TB Isle Resort, L.P., *865 F. Supp. 2d. 1172 (S.D.F.L. 2011)* (Fairmont v. Turnberry)

The facts of *Fairmont v. Turnberry* involved a relationship between a manager and owner that originated with the owner's original purchase of the hotel. The original owner had a sour opinion regarding the potential purchasers and refused to directly sell to them. As a result, Fairmont, the future manager of the resort, acted as an intermediary between the original owner and the future owner of the resort. The original owner sold the hotel to Fairmont, who had executed a separate agreement with the future owner, guaranteeing that they would purchase it and install Fairmont as the manager for 25 years with options to extend, creating a total 50-year contractual interest in the agreement. Because of the unique circumstances of the hotel acquisition, Fairmont also acquired equity rights in and approval rights over future projects. Future developments on the land required Fairmont's approval, and the developments had to be consistent with Fairmont's operations and policies. Moreover, Fairmont also secured the right to manage and brand such developments, and the right to invest in any development in the resort. As an additional layer of security, Fairmont also secured rights of first offer and first refusal in the event owner elected to sell the property.

Eventually, the owner sought to terminate the hotel management agreement because of alleged performance issues. The owner did not send Fairmont a notice of default or provide Fairmont with an opportunity to cure. One morning, the owner initiated a takeover of the hotel. A representative of the owner demanded the presence of Fairmont's senior management, and then informed them that they were de-branding the hotel. Private security then escorted Fairmont's management personnel from the premises, and the owner subsequently sent a termination letter to Fairmont. At the hearing that followed shortly after the events, Fairmont's witnesses claimed that owner's personnel carried a gun and their representative disconnected a phone connection during an outgoing call. The court characterized this takeover as "an orchestrated plot to take over the hotel without compliance with the ... HMA," and Fairmont as "a wronged company," which has been victimized by the resort owner and its principals.

Like the manager in *Woolley v. Embassy Suites*, Fairmont sought a preliminary injunction to force the reinstatement of Fairmont as manager pending the resolution of an arbitration of the dispute between owner and manager. Unlike *Woolley v. Embassy Suites*, Fairmont had been physically removed from the property without notice, and owner acted in the absence of a court order. The court applied New York law, and determined that Fairmont could only obtain an injunction if the hotel management agreement created an agency coupled with an interest. In addition to the plain text of the contract, which stated that it was an agency coupled with an interest, Fairmont asserted the following interests: (i) rights of first offer and refusal in the sale of the hotel, (ii) a strategic alliance agreement granting Fairmont equity rights in future developments, (iii) the right of possession and quiet enjoyment, (iv) the execution of the hotel management agreement as a condition to the initial sale of the property to the owner, and (v) a Fairmont affiliate's option to purchase the hotel.

Relying in part on *Woolley v. Embassy Suites*, the court held that several of these interests were insufficient because they were not "vested" rights. In particular, a vested right may not be one that is contingent. Under New York law, options are regarded as contingent because they depend on the happening of an event, the holder's choice. Rights of first offer and refusal ripen into options on an additional precedent event, the desire of another party to sell, and therefore are also contingent. The court dismissed the equity rights in future developments established in the strategic alliance agreement for the same reason; they depended on events that in fact never occurred. Finally, Fairmont argued that its affiliate's prior option to purchase the hotel constituted a sufficient interest. Not only is an option an unvested interest, but Fairmont's corporate separateness violated the unity principle discussed in *Pacific Landmark*.

Fairmont's other interests were insufficient because they were either dependent on a characterization of the hotel management agreement precluded by its terms, asserted an unrecognized interest, or violated the principle of corporate separateness (and the requirement that the right be vested). Fairmont's claim that it had the right of possession and quiet enjoyment was grounded on traditional landlord–tenant law, requiring a lease. The hotel management agreement declared it could not be construed as a lease. The execution of the hotel management agreement as a condition to the sale failed because neither Fairmont nor its affiliates had an ownership interest in the hotel.

Summary

Managers have encountered various barriers to gaining injunctive relief in the face of an Owner termination, primarily because their contracts had been characterized as revocable under agency theory. In these cases, the courts have regarded the following interests as insufficient to create an agency coupled with an interest:

- An affiliate entity's ownership interest in the hotel and security instruments encumbering the hotel for outstanding loans because the interest was not vested in the same legal entity.
- Brand name, reputation interests, and the contribution of confidential, proprietary, and trade secret information, because they are not interests in the hotels themselves, but incidental to the performance of management.
- Monetary consideration and contractual consideration for the initial sale of the hotel.
- Clauses specifically designed to prevent ouster or termination in the event ownership of the hotel changes because they only evidence the intent of the agency relationship and not its substance.
- Capital investments through a reduction in fees or tenant interests granted by leases because hotel management agreements preclude characterization as a joint venture, partnership, lease, or other relationship.

- Aggregate interests created by a series of contracts that are individually insufficient to create an interest in the subject matter of the agency.
- Rights of first offer and refusal, equity rights in future projects, and options because they are not vested interests.

Personal services contracts, the aftermath of Eden Roc and RC/PB v. Ritz-Carlton

In 2013, the *RC/PB* and *Eden Roc* cases ended the debate, at least for today, around what sorts of interests could give rise to an agency coupled with an interest under Florida and New York law respectively by holding that hotel management agreements are personal service contracts, and therefore not susceptible to enforcement through injunctive relief, such as when the manager is seeking to preserve its agreement, whether on the merits or in preliminary proceedings. These holdings have likely foreclosed hotel management companies' availability of injunctive relief.

According to the Second Restatement (Contracts) whether a personal services contract exists depends on the policy considerations that drive the ban on specific performance, namely, the 13th Amendment's prohibition against slavery and judicial economy. The Restatement posits that the contract must be non-delegable. Other relevant factors include:

- The degree of trust and confidence between the parties.
- The difficulty of objectively evaluating the performance of the contract.
- The length of time required for performance.

The intimacy of the relationship between the two parties in a personal services contract is similar to that of the agency relationship between master and servant, but under New York law, lacking the degree of control in the latter. The greater the degree of trust between the parties, difficulty in objective performance, and length of time all militate towards a finding that a contract is for the provision of personal services. The operation of a hotel requires a large degree of trust because the rights ceded to the manager are expansive, and the term of hotel management agreements usually exceeds several decades. Other examples often cited include contracts of actors, artists, sales agents, teachers, mechanics, cooks, and managers, *including* property managers.

Both *RC/PB v. Ritz-Carlton* and *Eden Roc* relied extensively on the degree of control and discretion passed on to hotel managers as it related to the difficulty of objectively measuring the performance of a hotel management agreement in holding that hotel management agreements were personal services contracts.

Marriott International Inc. v. Eden Roc (Eden Roc)

In *Eden Roc*, a fact pattern similar to the *Fairmont Turnberry* case arose. Renaissance (a brand owned and controlled by Marriott International), the manager, claimed that Eden Roc "stormed the hotel" with a security detail. Unlike *Fairmont v. Turnberry*, Renaissance also claimed that Eden Roc attempted to steal proprietary and confidential information. Renaissance contacted the police, who restrained Eden Roc from seizing the property requiring that they produce a court order before removing Renaissance from the property. However, here the lower court issued a temporary restraining order barring Eden Roc from removing Renaissance from the hotel, which Eden Roc appealed.

Eden Roc argued that the hotel management agreement was a personal services contract, and therefore temporary restraining orders or preliminary injunctions were unavailable as preliminary remedies because specific performance would not be available as a final remedy. Relying on *Woolley v. Embassy Suites* and *Fairmont v. Turnberry* as specifically analogous cases, Eden Roc argued that an entry of specific performance would violate the rule against providing injunctive relief for personal services contracts. Renaissance relied on the theory that the management agreement constituted a "complex business arrangement … which includes rights in real property" made between two sophisticated entities that expressly provided for specific performance as a remedy, and characterized Renaissance as an independent contractor.

The asserted property rights were familiar. They included a contractually granted right to quiet enjoyment of the hotel, clauses in the SNDA clause requiring a new owner to accept Renaissance's management of the hotel, investments and renovations made in the hotel, and reduced management fees. However, unlike previous managers, Renaissance avoided the *Woolley* line of cases by arguing that an agency relationship did not exist; in opposition, Eden Roc asserted an agency relationship. Instead, Renaissance argued that overall this was not a personal services contract, but a complex business relationship that had specific terms defining what rights and remedies were available to each party. Moreover, it contended that the agreement did not run afoul of the 13th Amendment because it merely required a prohibitive injunction.

In characterizing Renaissance's role as an independent contractor, Renaissance emphasized the limited nature of Eden Roc's control over Renaissance, undercutting the alleged master–servant component of a personal services contract. Relying on the policy behind the 13th Amendment, Renaissance concluded that a sophisticated entity such as Eden Roc could not reasonably be threatened with involuntary servitude. Regarding the policy of judicial economy, it asserted that a court would not have to heavily police a prohibitive injunction.

In a brief opinion, the Court in New York held that hotel management agreements were personal services contracts unenforceable by specific enforcement. It exclusively relied on the degree of discretion given to a manager in its holding. Analogizing to *Woolley v. Embassy Suites* and property management agreements, which had been held to be personal services contracts, the court ruled that Renaissance's performance could not be objectively measured because it had full discretion in the operation of the hotel. Therefore, it was a personal services contract. In dicta, it also noted that the agreement was not an agency agreement because of the owner's lack of control that corresponded with the manager's full discretion. Under Florida law, *RC/PB v. Ritz-Carlton* reached a similar conclusion, but did not foreclose the possible exception that the relationship could be an agency coupled with an interest.

RC/PB v. Ritz-Carlton

RC/PB v. Ritz-Carlton involved a heated dispute over corporate fees. RC/PB, the owner of the hotel, asserted that the parties had an agency relationship or fiduciary relationship and that Ritz-Carlton and Marriott, the managers, had operated the hotel for its own benefit. The hotel was deeply integrated into Marriott's overall global system, including the reservations system, the selection of key management, collective agreements entered into on behalf of the owner, insurance programs, centralized accounting, and global marketing schemes. Moreover, Ritz-Carlton and Marriott had invested extensively in the hotel's renovation, as managers have done in other cases. In particular, RC/PB alleged that Ritz-Carlton had taken advantage of its fees and inappropriately directed profits from RC/PB to Ritz-Carlton. This redirection of profits allegedly occurred in part by directing certain purchase and services contracts to an affiliate of Ritz-Carlton through a kickback scheme designed to conceal RC/PB's actual profits. Ritz-Carlton did not release these contracts to RC/PB for review, giving rise to its concerns about the improper funneling of funds. Moreover, RC/PB disputed the managers' tabulation of corporate charges, arguing they were excessive, prohibited by the hotel management agreement, and deprived the RC/PB of any meaningful profit from the operation of the hotel. As relief, RC/PB requested an order allowing it to terminate the hotel management agreement.

The Fifteenth Judicial Circuit of Florida, relying on *Woolley v. Embassy Suites, Hyatt v. Skopbank, and Eden Roc,* held that the hotel management agreement constituted a personal services contract, and emphasized the degree of control passed to the manager. Ritz-Carlton determined employment policies, conducted property maintenance, had purchasing authority granted by the owner, and could sue in the name of RC/PB. The hotel management agreement connected the exercise of Ritz-Carlton's discretion with its recognized expertise, which "undisputedly called for the rendition of services which require the exercise of special skill and judgment."

Ritz-Carlton presented several defenses. It argued that the contracts were assignable, and therefore could not be personal services contracts. However, the court ruled that this did not strip the hotel management agreement of its nature as a personal services contract because it was expressly provided for in the contract. As a separate defense, Ritz-Carlton also asserted that the performance termination provision allowed for performance to be objectively measured. The court relied on *Skopbank v.* Hyatt and *Eden* in rejecting this defense. In contrast to *Eden Roc,* its holding allowed for the viability of the agency coupled with an interest theory, calling it a "recognized exception" to the "well-settled law" law that specific performance is unavailable in disputes concerning personal services contracts.

Conclusion

These recent rulings have brought a degree of clarity to the legal theories available to hotel management companies that are faced with hostile removals under New York and Florida law. The agency argument, while not directly addressed in *RC/PB,* has been effectively disposed of by *Eden Roc.* All hotel management agreements, including the model, require that extensive control be transferred to the manager. (See companion website at www.routledge.com/cw/migdal to view the model form.) This discretion is exercised in strategic planning, integration into marketing and booking networks, day-to-day maintenance and operation of facilities, rate setting, daily expenditures, and selecting, training and overseeing employees. *Eden Roc* implicitly saw a temporary injunction as not merely requiring a prohibitive supervision, but the continuing evaluation of the manager's performance. The asserted interests that supposedly distinguish a personal services contract from a hotel management agreement are irrelevant in light of how hotel management agreements are structured. They do not create leasehold interests, partnerships, joint ventures, or other interests that differentiate them from personal services contracts. Under Florida and New York law, hotel management agreements are personal service agreements because:

- While hotel management agreements may be assignable, if such assignment rights are expressed in the agreement, they are not a bar to a finding that a hotel management agreement is personal services contract.
- The trust necessary for the performance of a hotel management agreement is within the degree of trust associated with personal services contracts because of the owner's cessation of rights on its own property.
- The hotel manager must exercise a wide range of discretion through the performance of the agreement.

Having mentioned the laws of certain states with respect to these critical cases, the laws of the State of Maryland are worthy of mention. The laws of the State of Maryland include the "Operating Agreements Act," codified at Md. Code § 23-102. Maryland has become a significant jurisdiction in the area of hospitality law because it is home to a number of significant hotel operating companies and hotel owners.

The statute provides that:

> [i]f a conflict exists between the express terms and conditions of an operating agreement and the terms and conditions implied by the law governing the relationship between a principal and agent, the express terms and conditions of the agreement shall govern. A court may order the remedy of specific performance for anticipatory or actual breach or attempted or actual termination of an operating agreement notwithstanding the existence of an agency relationship between the parties to the operating agreement (internal punctuation deleted).

Although the law is yet to be judicially addressed, particularly following the rulings already discussed with respect to personal service contracts and injunctions, it remains a unique statute with direct applicability to hotel management agreements. As already noted, *Woolley* and all the cases that followed noted the nature of Hotel Management Agreements as personal services contracts, not enforceable by specific performance. If, indeed, this remedial bar is rooted in the 13th Amendment, any legislative attempts to override it might not withstand scrutiny.

Injunctive relief and damages

Independent of the categorical bars to injunctive relief based on the nature of the relationship between an owner and manager of a hotel, obtaining injunctive relief also requires that the party seeking the injunction would suffer irreparable harm if the injunction is not granted, and that the balance of hardships weighs in favor of the party seeking the injunction. Each of these requirements has proven problematic for hotel management companies. However, while owners may have the power to terminate a hotel management agreement and remove the manager from their property, they do not necessarily have the right, giving rise to damages. Such damages may be extensive depending on the facts of any particular case.

The irreparable harm prong depends on whether the damages are calculable, and the moving party must make a clear showing of substantial, actual, and imminent harm.

Courts have routinely held that damage to goodwill and reputation, while difficult to ascertain, can be established through expert testimony. The primary consideration for a manager in a hotel management agreement is the management fee, which is calculable as well. Moreover, many hotel management agreements include provisions for damages calculations in the event of default, as well as liquidated damages. Finally, the "havoc in the industry" theory has failed; while such damage may be incalculable, it is overly speculative and hypothetical.

The balance of hardships has been emphasized less, but deserves some attention. *Woolley v. Embassy Suites* and *Fairmont v. Turnberry* both noted that the manager's primary loss would be monetary in nature. In contrast, an owner stands to lose significant property rights by enduring a continuous trespass or an unwanted agency relationship. These property rights cannot be cured purely with damages, for example, injunctions prohibiting future trespass are routinely granted by courts. A hotel manager's significant capital improvements in a hotel do not override these interests.

Although injunctive relief with respect to preserving a hotel management is unavailable, managers may avail themselves of it in more limited circumstances. *RC/PB v. Ritz-Carlton* has made it clear that managers can resist physical and hostile takeovers by contacting the local police. To protect their trade secrets, proprietary information, and other intellectual property they may seek an injunction fashioned to allow for the secure and court monitored transmission of these materials. An injunction could also be sought to prohibit an owner from causing brand or guest confusion through the use of the manager's name, or its intentional concealment of a transfer of the operations of the hotel. Injunctions must be proportionate, and if tailored in such a way under the right circumstances, might be more likely to be granted.

Finally, managers may be entitled to significant damages. Their fees are readily calculable; however other forms of compensation may be available depending on the hotel management agreement. Where considerations regarding brand name and reputation arise, damages can be established through liquidated damages or expert testimony. Certain statutory damages and punitive damages may be available against particularly bad actors. Intentional intellectual property infringements carry with them varied civil penalties as well. Finally, punitive damages are sometimes available for fraud, misrepresentation, or other significant wrongdoing. Hotel owners have the power, not the right, to terminate a hotel management agreement, and owners are answerable to the manager for its damages.

Key holdings

- Under Florida and New York law, hotel management agreements are regarded as personal services contracts lacking an agency component. Therefore, specific performance is not available to the manager as a remedy. As a consequence, when an owner terminates a manager's hotel management agreement, the owner may pursue an injunction removing the manager from the premises. The owner will be liable for any contract damages arising from a potential breach.
- Under California law, hotel management agreements are regarded as agency agreements. The California courts have not addressed a case leading to a ruling that a hotel management agreement was an agency coupled with an interest, and therefore, irrevocable. Outside California, in instances where courts have examined the availability of injunctive relief to the manager of a hotel, they have routinely commented

that it would not be available because the particular agency relationship was not coupled with an interest. We do not know if an agency coupled with an interest would nevertheless be considered a personal services contract.

- In each jurisdiction, even if a hotel management agreement created an agency coupled with an interest enforceable by specific performance or injunction, an injunction would not be granted because the damages arising from the breach are ascertainable based on hotel management fees and other fixed contractual remunerations, expert testimony, and other means. The damage to the manager is regarded as reparable because it can be ascertained by examining the contract and evidence specific to each case, and accordingly injunctive relief is unavailable.
- Contractual provisions that explicitly provide for injunctive relief as a final or preliminary remedy do not overcome the traditional bar against the specific performance of personal services contracts. Even if the provision in question further provides that the parties are sophisticated investors, have had their attorneys review the contract, and that the remedy was specifically bargained for, a court may still refuse to grant an injunction because it runs afoul of the 13th Amendment.
- While owners retain the power to terminate a hotel management agreement and remove the manager from the property, they remain liable for damages resulting from any potential breach of contract. Damages may include liquidated damages, actual damages as established by the hotel management agreement through the management fee and other compensation, consequential damages arising from loss of reputation, infringement of intellectual property rights, or contractual interference. Because hotel management agreements are typically secured for extended periods of time, these damages may be significant depending on the facts of a particular case.

Cases cited

FHR TB, LLC v. TB Isle Resort, LP., 865 F. Supp. 2d 1172 (S.D.F.L. 2011)

Government Guarantee Fund of the Republic of Finland v. 35 Acres Associates, 95 F.3d 291 (3d Cir. 1996)

Marriott International, Inc. v. Eden Roc, LLLP., 962 N.Y.S. 2d 111 (N.Y. App. Div. 2013)

Order on Motions to Dismiss of Defendants, *RC/PB, Inc. v. Ritz-Carlton Hotel Company, L.L.C.*, No. 502011-CA-010071 (Fla. 15th Cir. Ct.)

Pacific Landmark Hotel, Ltd. v. Marriott Hotels, Inc., 23 Cal. Rptr. 2d 255 (Cal. Ct. App 1993)

Title 23, Operating Agreements – Hotels and Retirement Communities, MD Code Ann., Comm. Law § 23-101-106 (West 2014)

Restatement (Second) of Agency §§ 138, 139 (1958)

Restatement (Third) of Agency § 3.12 (2006)

Woolley v. Embassy Suites, Inc., 227 Cal. App. 3d 1520 (Cal. Ct. App. 1991)

5 Franchise Agreement Basics

A Franchise Agreement essentially provides Owner with the benefits of access and rights to the Intellectual Property associated with the Brand and the Brand's System, a global reservation system, loyalty program, and an established guest base in exchange for license fees, marketing fees, and reservation fees. Franchisee will either self-manage the Hotel or hire a third-party management company to manage and operate the Hotel in accordance with Brand System and Standards, the Franchise Agreement and the Management Agreement between Owner and Manager. On occasion, Franchisor or its affiliate will manage the Hotel pursuant to a separate Management Agreement.

With a Franchise Agreement in place, Franchisor is removed from day-to-day operations of the Hotel and the burden is placed on Franchisee to either manage the Hotel or to hire a Third-Party Manager approved by Franchisor in advance and to ensure the operation of the Hotel is in conformity with the Franchise Agreement and Franchisor's Brand System and Standards.

In the United States, including its territories and possessions, the offer and sale of a Franchise is governed by the Federal Trade Commission ("FTC") pursuant to the FTC's Franchise Rule (the "FTC Franchise Rule"). Under the FTC Franchise Rule, if the transaction does not qualify for an exemption (see discussion that follows), a Franchisor is required to prepare a Franchise Disclosure Document ("FDD") and deliver it to a representative of the prospective Franchisee entity at least 14 days before the "sale" of a Franchise.

Franchising is also regulated at the state level. In addition to the FTC Franchise Rule, 15 states have Franchise registration/disclosure statutes (the "Franchise Registration States"), which require some form of registration and disclosure prior to the offer and sale of a Franchise. In addition, 23 states have Franchise relationship laws which, as the name suggests, govern the relationship between Franchisors and Franchisees after the sale of a Franchise.

Each of the Franchise Registration States, like the FTC Franchise Rule, requires presale disclosure of an FDD to a prospective Franchisee. The FDD is a comprehensive document, consisting of 23 separate items, designed to provide the prospective Franchisee with all of the material information that it needs to make an informed business decision. It provides critical information regarding Franchisor, its parents or affiliates; "control people" (i.e., those individuals who serve as directors or officers or who have management responsibility relating to the Franchised business or the offer and sale of the Franchise). The FDD also includes important information concerning the investment costs and fees that will be payable to Franchisor or its Affiliates. It also sets forth, in plain English, the material terms of the relationship between Franchisor and Franchisee. Most importantly, the FDD includes copies of all agreements (including the Franchise Agreement) that Franchisee will be required to sign. Some, but not all, guest lodging FDDs also contain a form of Lender Comfort Letter. (See the companion website. Please visit www.routledge.com/cw/migdal to view the examples for current base form Franchise Agreements from Marriott International, Inc., and from Hyatt Hotels Corporation.)

It is worth noting that the FTC Franchise Rule (and, to a lesser extent, the laws in the Franchise Registration States) contains many exemptions, which, if all elements of the exemption are met, render the FTC Franchise Rule inapplicable to the transaction. In other words, if the transaction is exempt, Franchisor has no obligation to provide pre-sale disclosure to the prospect and, more importantly, Franchisor is not bound by any of the requirements of the FTC Franchise Rule, including, for example, its onerous FDD delivery timing requirements and the prohibition against supplying financial performance information not otherwise included in the FDD.

Exemptions to the FTC Franchise Rule are particularly helpful in guest lodging since most guest lodging transactions qualify for one of the available exemptions. If the transaction is exempt, Franchisors need not comply with the requirements of the FTC Franchise Rule. Examples of exemptions which may apply to a guest lodging transaction include the following.

Large investment exemption

Section 436.8(a)(5) of the FTC Rule exempts from its pre-sale disclosure requirements the sale of a Franchise to an investor where the total initial investment (excluding financing from Franchisor) exceeds $1,084,900 and Franchisee acknowledges in writing that the initial investment will in fact exceed $1,084,900.

According to the Statement of Basis and Purpose which accompanied the FTC Franchise Rule (the "SBP"), "the basis for the large investment exemption is not that 'sophisticated' investors do not need pre-sale disclosure, but that they will demand and obtain material information with which to make an informed investment decision regardless of the application of the [FTC Franchise] Rule." The large investment should be calculated by reviewing the prospective Franchisee's initial investment as would be set forth in Item 7 of the FDD prepared in accordance with the FTC Franchise Rule to determine whether the exemption has been met.

In addition, the required $1 million investment necessary for a Franchisor to take advantage of the FTC Franchise Rule's exemption "need not be limited to a single unit … A multi-unit Franchisee investing the threshold amount (or more) in a number of units is just as sophisticated as another Franchisee investing a like amount in a single unit."

Critically, under the FTC Franchise Rule, the value of assets that are the subject of Franchise conversion or transfer transactions will count when computing the required threshold. For example, a Hotel Franchisee that is converting from one Franchised brand to another can take into account the initial investment in the original Brand (including all improvements to the Property) when determining whether the elements of this exemption have been satisfied.

It is important to note, however, that the application of the large investment, investment exemption is more difficult when Franchisee entity is comprised of multiple investors. In that case, the exemption will apply only if at least one individual in a Franchisee investor group invests enough money to meet the required level of investment.

Sophisticated Franchisee Exemption

In addition to the large investment exemption, the FTC Franchise Rule also recognizes an exemption based on the net worth and sophistication of Franchisee, which is frequently referred to as the "Sophisticated Franchisee Exemption" or "large franchisee exemption." This exemption applies to entities (including any parent or Affiliates) that have been in business for at least five years and have a net worth of at least $5 million. A large Franchisee need not have five years of business experience in franchising or in the industry that Franchisee will enter as a result of the Franchise; five years of business experience in any business will suffice.

The type of prospective Franchisee entities eligible for the "Sophisticated Franchisee Exemption" include corporations, partnerships, other business entities and individuals. With regard to individuals, the SBP observes: "Nothing prevents an 'entity' under this provision from being an individual, but most individuals who have been in business for at least five years and have generated an individual net worth of at least $5,424,500 are likely to have created a corporation or other formal organization through which to conduct business."

California, Maryland, South Dakota, and Wisconsin exempt offers and sales of Franchises to "sophisticated franchisees" from the registration and disclosure provisions of each state's law.

Fractional Franchise Exemption

The FTC Franchise Rule defines a "Fractional Franchise" as "a Franchise relationship that satisfies the following criteria when the relationship is created: (1) Franchisee, any of Franchisee's current directors or officers, or any current directors or officers of a parent or Affiliate, has more than two years' experience in the same type of business; and (2) parties have a reasonable basis to anticipate that the sales arising from the relationship will not exceed 20% of Franchisee's total dollar volume in sales during the first year of operation." "Fractional Franchisees" are exempt from the disclosure requirements of the FTC Franchise Rule.

California, Illinois, Minnesota, New York, South Dakota, and Wisconsin also exempt the offer and sale of "Fractional Franchises" from the pre-sale registration requirements.

Hotel Franchise Agreements are difficult to negotiate and modify. Most Franchisors are unwilling to heavily negotiate a Franchise Agreement and will not make revisions because Franchisor must maintain a uniform Franchise System, where all Franchisees are generally operating under Franchise Agreements with the same material terms. Most importantly, Franchisor must ensure that its Franchise Agreement affords it the ability to require its Franchisees to operate the Hotel in accordance with Franchisor's Brand System and Standards, which may be modified from time to time throughout the lengthy Term of the Franchise Agreement. Understanding this practical reality, it is a better practice to realize that Franchise Agreements are typically only modified with respect to a few key areas, and then focus on those key areas in negotiating the terms of the Franchise Agreement with Franchisor.

Some of the common key components of the Franchise Agreement that may be worthy of negotiation include:

- The Property Improvement Plan ("PIP") (i.e., the extent of renovations at the Hotel that Franchisee will be required to undertake either prior to or periodically during the term of the Franchise Agreement).
- Area of Protection/Exclusivity (i.e., the "territory" surrounding the Hotel in which Franchisor will agree to not place a competitive Hotel).
- Renewal Periods and at which party's option (in guest lodging, "renewal" is referred to a "Re-Licensing").
- FF&E Reserve amounts.
- Ramp-up periods for the various fees.
- Minimum marketing spend from Franchisor's Marketing Fund.
- Franchisee transfers.

The Franchise relationship and the limited willingness to negotiate revisions assure Franchisor a great deal of control over the Brand and its reputation. This is not necessarily bad for Owner. It provides Owner with a degree of comfort that Franchisor will protect the consistency and integrity of the Brand, which is invaluable in the Franchise context. A guest that suffers a bad experience in one Franchised Hotel may hold it against the entire Brand for years to come. Each individual Franchisee has an interest in knowing that every Franchisee must adhere to the same Brand Standards, and has executed a Franchise Agreement that is essentially the same as the Franchise Agreement executed by any one Franchisee.

Provisions for random quality assurance testing for customer care and service, guest satisfaction and other similar performance verification are addressed throughout the Franchise Agreement and they afford Franchisor the ability to ensure quality control over its Brand, which takes the place of the more intrusive management where Franchisor is a Brand Manager. Additionally, if Franchisee does not desire to or is unable to manage the day-to-day operation of the Hotel, virtually all Franchise Agreements require the approval of Franchisor of any Third-Party Manager or any change of the Third-Party Manager. Under certain circumstances, Franchisor or one of its Affiliates will manage the Hotel under a separate Management Agreement.

The Property Improvement Plan or "PIP" is the vehicle through which Franchisors ensure that the Hotel remains in compliance with the Brand System and Brand Standards. When a Franchised Hotel changes hands or Brand Standards change, Franchisor will issue its PIP to Franchisee and Franchisor and Franchisee will negotiate schedules and guidelines for Franchisee's compliance. Compliance with a PIP can be an expensive proposition for Franchisee. Although Franchisee must ultimately comply with the PIP, it is common for Franchisees and Franchisors to segregate the PIP into elements for which immediate and complete compliance is mandated, such as life safety matters, and other elements that may be more aesthetic or visual, for which compliance can be phased in and may span more than one Operating Year. Purchasers of existing Hotel assets and those converting from one Brand to another should anticipate a PIP and accommodate the cost in the initial budgets.

The basics of franchise agreements include the following.

Grant of the Franchise

Franchisor will grant Franchisee a non-exclusive, non-transferable and non-sub-licensable license for the marks and trade name related to the Brand. The grant of the Franchise is for the specific Hotel at the specific location identified in the application initially submitted by Franchisee. Franchise Agreements are personal in nature to Franchisee as identified in the application and not subject to assignment to a successor Franchisee, unless Franchisor consents. This consent may be granted after the successor Franchisee has completed a detailed application for a Franchise and paid the required application fee. The application form and the application fees are described in the FDD.

Name of the hotel

The names that Franchisee can use at the Hotel and in advertising referring to the Hotel will be set forth in the Franchise Agreement. Deviations from what has been approved and included in the Franchise Agreement are not permitted without Franchisor's consent, which will be in Franchisor's sole discretion.

Term and renewal term

The term of the Franchise Agreement will be negotiated and Franchisee should take care to coordinate the Franchise Agreement Term with the Term of the Management Agreement of Franchisee's Third-Party Manager so that the term of the Management Agreement is no longer than the term of the Franchise Agreement, although it may be shorter. A typical Franchise Agreement Operating Term can be anywhere from 10 to 25 years depending on a number of factors, such as desirability of the location, the extent of the start-up costs and length of stabilization period of the Hotel. An initial period of 20 years is common for good locations, and often consistent with Franchisee's business plan to stabilize Hotel operations and become profitable. Franchise Agreements will not necessarily include a renewal term, and this will be evaluated case-by-case. A Franchisee may wish to reevaluate its options after the initial term. Of course, renewal terms with a continuing Area of Protection can be of significant value to Franchisee of a well-located franchised Hotel. Renewal Terms can include multiple renewals. Most often, each Renewal Term is five years. The right to exercise the Renewal Option generally belongs to Franchisor so long as Franchisee is not in default under the terms of the Franchise Agreement, but this is something Franchisee can negotiate as neither party may want to be bound to continue the Franchise at the Hotel.

Franchisor will assess and charge Franchisee a number of fees. The various Franchisors active in the hospitality industry may use different terminology to identify their fees, but they will fall into the following general categories:

Types of fees

- Application Fee;
- Start-up Administration Fee;
- Royalty Fee;
- Franchise Fee;

- Technical Services Fee (for new construction or material renovation)
- Reservation Fee
- IT Fee;
- Electronic System Fee
- Marketing Fee.

Fees are defined as follows:

- The Application Fee is the fee assessed by Franchisor to obtain the Franchise in the first instance. The Application Fees vary by Franchisor, but this is public information available from most Franchisors, as well as industry publications. Generally, the Application Fee is based is based on a flat rate, modified to account for the number of approved Guest Rooms at the Hotel.
- Start-Up Administrative Fees are intended to cover administrative services such as training, approval of marketing and design materials, purchasing, staging, programming and installation services, and costs of manuals.
- The Royalty Fee is a percentage of Gross Room Revenues and is payable to Franchisor on a monthly basis during the Term of the Franchise Agreement. Note the variation from the management context where Manager's Base Fee is calculated as a percentage of Gross Revenues of the Hotel rather than the more limited Gross Room Revenues. Note also that if the Hotel is a full service Hotel, Gross Food and Beverage Revenues might also be included in the calculation of the Royalty Fee.
- The Marketing Fee is also a percentage of Gross Room Revenues and is payable to Franchisor on a monthly basis during the Term of the Franchise Agreement as a contribution to the Brand's Marketing Fund.
- The Reservation Fee can be expressed in a variety of ways including a fixed charge per reservation or a percentage of the Gross Room Revenues.
- The IT Fee can include a computer systems software license fee or other similar charges and can vary between fixed charges per month to a percentage of Gross Room Revenues.

Franchisor's out-of-pocket expenses incurred in connection with the Franchise Agreement, including any executive and staff time involved in the enforcement of the Franchise Agreement or marketing costs, will be subject to reimbursement by Franchisee.

All sales and marketing materials related to the Hotel developed by Franchisee and the method of use of any of the marks granted under the Franchise Agreement must be provided to Franchisor for review and require the written consent of Franchisor. Franchisor has sole and absolute discretion over the use of the trademark and Intellectual Property at all times and maintains ownership over all Intellectual Property.

The Franchise Agreement provides for very strict compliance by Franchisee. Any violation will require Franchisee to take all actions to immediately cease and desist from the behavior, take actions requested by Franchisor to remedy any violations and comply with the marketing standards all at the sole cost of Franchisee. Often just a few violations of the Brand Standards during any 12-month period can result in a default under the Franchise Agreement and Franchisor having the right to terminate the Franchise Agreement. In addition, most Franchise Agreements also provide for interim remedies (short of termination) that a Franchisor can implement if a Franchisee is in default. Examples of interim remedies include imposition of a "non-compliance fee" (typically 1% of Gross Rooms Revenues) or removal of Franchisee from the Brand's reservation system until the Hotel is in compliance with its obligations under the Franchise Agreement.

Management and operation of the Hotel can be through Franchisee, if approved by Franchisor, or through a professional Hotel Manager. The common general criteria for Franchisor's approval of Franchisee's Manager will include the following elements:

- Manager must be deemed qualified by Franchisor;
- Manager and Franchisee would deliver to Franchisor an acknowledgement to be countersigned by Franchisor, and Franchisors will refuse to grant consent unless Manager can satisfy them that Manager:
  - possesses sufficient financial capacity;
  - is sufficiently experienced in the exercise of managerial skills or operational capacity; and
  - is capable of discharging Franchisee's obligations and requirements of the Franchise Agreement.

Franchisors will require information about Manager and even access to other businesses that Manager operates. Franchisor will evaluate not only Manager, but the Management Agreement as well. Franchisor's evaluation of the Management Agreement is generally limited to determining that the Management Agreement is consistent with the terms of the Franchise Agreement and the acknowledgment.

In the event of a change in control of Manager or if Manager becomes a Competitor, or if there is a material adverse change to the financial status or operational capacity of Manager, Franchisee must promptly notify Franchisor and Franchisor has the right to require the termination of the Management Agreement and the replacement of Manager. Franchisor can also compel replacement of Manager if the Hotel performs poorly under Franchisor's quality assurance program. Franchisee's failure to replace Manager in timely fashion in that situation can mature into a Franchisee default under the Franchise Agreement.

When it comes to matters concerning the operation of the Hotel, Franchisor can communicate directly with Manager and need not go through Franchisee. Franchisor will have the right to enter the Hotel to conduct inspections to ensure that Franchisee is in compliance with the Franchise Agreement. Franchisor also retains the right to take photos and videos of the interior and exterior of the Property. Franchisor retains the right to engage mystery shoppers as a form of quality control. All costs associated with inspections are that of Franchisee.

Franchisor retains the option to terminate the Franchise Agreement in an event of default, including the option of seeking any other remedy available at law or equity to protect its marks and brand.

Under certain circumstances, restrictions on the building, licensing or granting of rights to other parties for the Franchise and/or management of a similar property under the same Brand by Manager may be granted for a limited period of time for the benefit of Franchisee.

It is the responsibility of Franchisee to oversee the Franchised business at all times. This includes ensuring that all Brand Standards are met and that the operations of the Hotel comply at all times with the Brand Standards, together with training by Franchisor for persons holding certain management positions. Franchisee bears the cost of the training, although it is mandated by Franchisor.

Franchisee should understand that Franchisor's Brand Standards are not static and will change during the Term of the Franchise Agreement. Franchisee acknowledges in the Franchise Agreement that when changes to the Brand Standards occur, Franchisee is required to comply with any changes. This can lead to unfortunate results for the unknowing or unprepared Franchisee. It was not long ago that Gross Revenues from Hotel operations were inadequate to cover Operating Expenses, taxes, and Debt Service. A change in Franchisor's Brand Standards could lead Franchisee to add compliance with revised Brand Standards to the list of things that Franchisee would not have the cash to cover, absent new debt or an additional investment of equity into the Hotel. If there are insufficient cash resources to comply with the new Brand Standards, Franchisee can seek a delay in implementation from Franchisor, but there is no assurance that Franchisor will grant the request. If Franchisor insists upon timely compliance and Franchisee is unable to comply, it is a default under the Franchise Agreement that can lead to loss of the Franchise as well liability for the Liquidated Damages under the Franchise Agreement. The default under the Franchise Agreement and loss of the Franchise will be an Event of Default under Franchisee's loan. If Franchisee cannot overcome the Loan default with its Lender, it can result in acceleration of the principal balance of the Loan and action by the Lender to exercise its rights and remedies under the loan documents, including Foreclosure. There is an important lesson to be learned about maintaining adequate reserves. Many experienced Franchisees appreciate and understand the value to a Franchisee in having Brand Standards, even though it comes with the cost of paying to maintain those Brand Standards during the entire Term of the Franchise Agreement, even as the Brand Standards change.

Franchisee is obligated to maintain, install, repair or substitute any FF&E as required under the Brand Standards. This includes signs, vending machines, communications systems, and supplies as may be required over the course of time. This is all at the sole expense of Franchisee.

An FF&E Reserve is typically required by Franchisor or Franchisee's Lender or Manager for capital or FF&E requirements. As with Management Agreements, the amount of the FF&E Reserve is expressed as a percentage of monthly Gross Revenues. Funds deposited to an FF&E Reserve represent money of Owner, so there is often a tension between holding back reserve funds and distributing cash to ownership. If funds are required for compliance with Brand Standards, capital expenditures or FF&E, that cannot be satisfied from reserve funds, then previously distributed cash may need to be recalled in the form of a capital call because all these expenses are the sole and exclusive obligation of Franchisee.

Franchisee is required to use Franchisor's global reservation system and its software in order to access the Brand. Once again, Franchisor reserves the right to change, alter or otherwise modify the System, as well as the right to require other systems to be used, at the sole cost of Franchisee.

Although Franchisor will have national marketing campaigns for the entire Brand System, Franchisee is responsible for its own local marketing, promotional, and public relations programs. These types of program require prior approval by Franchisor and must at all times comply with the Brand Standards.

National marketing campaigns coordinated by Franchisor are paid for by Franchisee through the contributions made by Franchisee to Franchisor's Marketing Fund.

The Marketing Fund is used for the benefit of the entire System and Brand rather than any particular region or any one Hotel. Monies in the Marketing Fund are not held in segregated accounts by property or by Franchisee, but instead are commingled with other marketing funds collected by Franchisor. There is no trustee or fiduciary duty owed to Franchisee with respect to the Marketing Fund. However, most Franchisors do prepare an accounting of Advertising Fund expenditures on an annual basis, which is available to Franchisees on request. It is incumbent on Franchisee to specifically raise and negotiate specific marketing efforts to benefit Franchisee's Hotel based on the relevant facts and circumstances, such as it being a new or newly renovated or converted Hotel that should receive certain special attention as it opens and stabilizes. If this is not negotiated in the Letter of Intent or the Franchise Agreement, Franchisee's request that Franchisor undertake programs that specifically target the Hotel for advertising will be determined entirely at the discretion of Franchisor.

Franchisee typically has the obligation under the Franchise Agreement and disclosures made in its FDD to keep full and adequate books of accounts and other records in accordance with generally accepted accounting principles and the Uniform System.

Unlike the Management Agreement context where Manager reports to Owner, under a Franchise Agreement, Franchisee must provide monthly financial statements to Franchisor. These monthly statements must reflect the financial results of the Hotel, including comparison reports, updates to the approved annual budget, and any other reports specified in the Franchise Agreement.

In addition to monthly reporting, Franchisee will provide annual unaudited statements certified by Franchisee as true and correct and subject to an audit by independent accountants, which is considered an operating expense of the Property.

Default and termination

The Franchise Agreement will include provisions for the non-defaulting party to terminate, usually on notice and an opportunity to cure, in the event of a material default by the other party. A material default will generally include the breach or failure of any party to comply with its covenants or agreements contained in the Franchise Agreement other than those that are considered minor, immaterial or insubstantial in nature. There are certain events that will give Franchisor a right to immediately terminate the Franchise Agreement with no opportunity to cure. These events usually include the following:

- insolvency of Franchisee or a Guarantor;
- voluntarily seeking bankruptcy protection by Franchisee or a Guarantor;
- adjudication of bankruptcy as to Franchisee or a Guarantor;
- levy of execution against the Hotel on a money judgment;
- foreclosure action under a mortgage;
- threat or danger to public health or safety;
- disclosure of Confidential Information;
- breach of material representations or warranties;
- repeated underreporting by Franchisee;
- acts of moral turpitude, such as conviction of a felony as to Franchisee, a Guarantor or someone holding a Controlling Interest;
- Franchisee or a Guarantor becomes a Competitor;

- Franchisee dissolves or loses the right to possession of the Hotel or the Hotel ceases operations
- improper transfer of the Franchise Agreement
- repeated performance failures under Franchisor's quality assurance or guest satisfaction program.

Franchisor will have an additional right to terminate the Franchise Agreement if Franchisee fails to make any capital expenditures or other repair expenditures necessary for the continued operation of the Hotel that are required in the reasonable opinion of Franchisor to maintain Brand Standards or the marketing standards.

Franchise Agreements differ from Management Agreements in the crucial area of termination fees or Liquidated Damages. Although some Management Agreements include a Liquidated Damages clause, such a clause is a common feature in a Franchise Agreement. The majority of Franchise Agreements include Liquidated Damages provisions that address at least three time periods: the period of time between the execution of the Franchise Agreement and the Opening Date of the Hotel, the period of time between the Opening Date of the Hotel and the end of a Stabilization Period, and the remainder of the Franchise Term. One very common configuration of a Liquidated Damages clause would state that if the Franchise Agreement terminates on or after the end of the negotiated Stabilization Period, such as the third anniversary of the Opening Date, the Liquidated Damages are the product of (i) the average monthly Royalty Fee and Marketing Fee that Franchisee was obligated to pay to Franchisor during the 12 full calendar month period before the month of termination, without regard for any provision in the Franchise Agreement deferring or reducing any portion of those fees, multiplied by (ii) the negotiated time period expressed in months, such as 36 to 60 or the number of months remaining in the Franchise Agreement Term, whichever is less. The Liquidated Damages are typically applicable in all circumstances except where Franchisor is the defaulting party. Additionally, a special circumstances damage provision is often provided for an increased multiplier in the event that the default or termination under the Franchise Agreement results in additional terminations of other Franchises held by Franchisee or its Affiliates within a 12-month period.

Care must be taken to look carefully at the Franchise Agreement, while the termination fee is expressed as Liquidated Damages, it is often not exclusive of any other remedies that may be available at law or in equity to Franchisor.

Transfers

The restrictions on transfers are fairly onerous for Franchisee, harking back to the concept that the granting of the Franchise by Franchisor is personal to Franchisee and based on the facts described in Franchisee's application, including Franchisee's ownership structure, and Franchisor's reliance on Franchisee to operate the Hotel in accordance with the Franchise Agreement.

Transfers within Franchisee's family or for estate planning purposes are generally not considered transfers that require Franchisor's consent. Most other transfers where Franchisee might bring a new partner into the venture with a change in control will be subject to Franchisor's consent.

Not only will Franchisee have to plan on a request for Franchisor's consent and allocate sufficient time to the process, most Franchise Agreements provide for fees for reviewing potential transfers, sale or other disposition of the Hotel.

The transfer of ownership interests or sale of the Hotel, with Franchisor's consent, will usually also trigger Franchisor's right to require a new PIP at the Hotel and it will also require the Transferee to execute Franchisor's then current Franchise Agreement. (Please visit www.routledge.com/cw/migdal to view the examples and for further analysis of the forms.)

Indemnification

The indemnification provisions of the Franchise Agreement run from Franchisee to Franchisor. Franchisee will indemnify, defend, and hold harmless Franchisor from and against all losses, costs, liabilities, damages, claims, and expenses of every kind and description, including allegations of negligence by Franchisor arising out of or resulting from almost everything that occurs at the Hotel. This would include the following:

- the unauthorized use of the Proprietary Marks
- the violation of Applicable Law
- the construction, renovation, upgrading, alteration, remodeling, repair, operation, ownership or use of the Hotel or of any other business conducted on, related to, or in connection with the Hotel.

In a Franchise relationship, Franchisee (or its approved Manager) is solely responsible for the operation of the Hotel. This is what leads to Franchisee's obligation to indemnify Franchisor for the broad range of matters just listed.

Franchisor's indemnification of Franchisee will be limited to acts arising out of or resulting from the permitted use of the brand name and marks during the Term of the Franchise Agreement.

Financing

Franchisees should anticipate that the form of Franchise Agreement will include limitations on Franchisee's financing of the Hotel. Common limitations include general language that the financing must be on commercially reasonable terms, as well as a specific debt service coverage ratio limitation. This will be accompanied by Franchisee's acknowledgement in the Franchise Agreement that Franchisor has real estate rights with respect to the Hotel, and the right to file an instrument among the applicable land records to memorialize that interest. This can catch Franchisees by surprise, and potentially cause an unintentional default under Franchisee's Loan Documents that often contain a covenant to Mortgagee that there are no other persons with any interests in the Hotel. These Franchisor rights will generally be subordinate to the lien and effect of Mortgagee's Mortgage.

The springing franchise

There may be situations or locations that begin as Hotels subject to management by the Brand, but if the Management Agreement terminates (sometimes regardless of the reason for the termination), it triggers a Franchise Agreement instead of loss of the location. In this situation, Owner and the Brand may negotiate a Management Agreement with complete management in the hands of the Brand as Manager in the typical sense. However, Owner may negotiate a termination right in the Management Agreement, subject to the Brand's right to continue to have a presence at the Hotel, but under a Franchise Agreement.

The relationship of the parties is dramatically altered, but the Hotel is retained in the Brand family nevertheless. This can be very effective when an Owner will develop multiple hotels with the Brand. In the early stages of the relationship the Brand may require management control to establish market presence and to control the guest experience. Then, once the Brand is satisfied with Owner's ability to self-manage or engage professional management, the Brand may permit the Management Agreement to be terminated in favor of a Franchise Agreement. Pre-negotiation of the Franchise Agreement becomes an important component of the transaction, subject to changes to the System in the period between the Opening Date of the Hotel under the Management Agreement and the effective date of the Franchise Agreement.

Guarantor

A unique feature of the Franchise relationship that often surprises Franchisees is that Franchisors commonly require a Guarantor to guarantee the performance of Franchisee's obligations under the Franchise Agreement. The Franchise Agreement will include a form of Guaranty to be signed and delivered by the Guarantor when Franchisee signs the Franchise Agreement. The Guarantor will be required to retain a minimum level of net worth and it will be a default under the Franchise Agreement for the Guarantor to be in default under the Guaranty, including the required minimum net worth. Franchisees have found it to be exceedingly difficult to negotiate the Guaranty out of the Franchise Agreement. The chances are slightly improved when Franchisee is also Owner of the Hotel because Franchisor knows it has access to the value of the Hotel itself, but even under these circumstances, the negotiation to remove the Guaranty can be unsuccessful. Franchisees have looked for alternative negotiation points in lieu of the complete removal of the Guaranty. Although each must be evaluated case by case, some common approaches include the following:

- Negotiate the Guarantor. Look to an Affiliate or related entity rather than an individual principal of Franchisee.
- Negotiate the minimum net worth. Connect the Liquidated Damages under the Franchise Agreement to this requirement.
- Negotiate a "cap" on the Guaranty.
- Negotiate a term on the Guaranty so that it will diminish over time or completely sunset.

Many Franchise Agreements also include riders and exhibits to address technology and computer licenses, and special provisions that Franchisor or Franchisee may have identified in the course of their negotiations. For example, a Hotel that is so well located that Franchisor considers the location to be of special strategic value to the Brand may lead Franchisor to negotiate for a strategic right of first refusal or right of first opportunity to purchase, that will be reflected in a rider.

6 Subordination Non-Disturbance and Attornment and Comfort Letters

The "SNDA" or Subordination, Non-Disturbance and Attornment Agreement is a common and familiar document in the hotel financing arena. Even with some of the more interesting transaction structures in the REIT environment with operating leases and a careful segregation of the ownership of the Hotel from the operation of the Hotel, there will be an instrument as part of Owner's debt financing for the Hotel intended to govern how the Hotel Owner, Hotel Manager and Owner's Mortgagee will behave in the event of the Hotel Owner's default under its loan instruments with the Mortgagee.

The form of the SNDA is often the first battleground of this debate. In the negotiation of the Hotel Management Agreement, Owner and Manager will often pre-negotiate the form of the SNDA and attach it as an exhibit to the Management Agreement or include within the body of the Management Agreement some of the elements that must be set forth in the SNDA. Both Owner and Manager might find it beneficial to not pre-negotiate the SNDA or its terms in the Management Agreement, which frequently will be executed long before Mortgagee is identified, and instead agree to enter into a commercially reasonable form of SNDA that will not increase the burdens or decrease the benefits afforded Owner and Manager in the Management Agreement.

The large national Brands will have well and fully developed and battle-tested SNDA forms that will be modified, if at all, only after considerable dialogue among Mortgagee, Hotel Owner and Hotel Manager. Similarly, the lending community will have its battle-tested forms of SNDA. These may be exhibits to the loan application or commitment letter from the Mortgagee, and frequently overlooked at that stage of the relationship of the Mortgagee and Owner, and later exhibits to the loan documents. Woe unto the less than careful Owner that finds itself with one required form of SNDA as an exhibit to the executed Management Agreement and a second required form of SNDA as an exhibit to the loan documents. Early recognition that there will be an SNDA requested or required by Manager, and attention to the forms or text can save significant time and money later in the negotiations.

The various forms of SNDA seen in the marketplace from time to time provide insight into the objectives and philosophy of some Mortgagees prevailing in that market climate, usually as a result of recent negative experiences, and where the industry is within the current market cycle. For example, some Manager-centric SNDA forms recite that they seek to provide for the continued management of the Hotel pursuant to the Management Agreement even in the face of Owner's default under its loan. Compare that to the forms that recite that the Mortgagee has required the execution and delivery of the SNDA to govern should the Mortgagee foreclose the lien of the security instrument or otherwise succeed to the rights of Owner. The objective always remains to have a clear and concise expression of the respective rights and obligations of the parties that will spring into action in the event of an Owner loan default and the exercise by the Mortgagee of its rights and remedies under the loan documents. The common challenge is that Manager wants to continue to manage pursuant to the Management Agreement as if Owner had not defaulted on its loan, and Mortgagee, now dealing with a defaulted loan, wants a clear and rapid exit, that may include the appointment of a receiver and the termination of Manager. These are not easily reconcilable positions.

Most SNDA forms include language that permits Owner and Manager to behave under the Management Agreement without regard to the SNDA so long as there is no event of default under the loan documents. Of course, depending upon the Mortgagee and its then current policies and procedures, it should not be a surprise to be confronted with a Mortgagee seeking Budget approval rights and any number of lockbox control agreements and cash collateral agreements even when there is no Owner loan default. This is consistent with the typical loan documentation that will include with the SNDA a collateral assignment to the Mortgagee of Owner's rights under the Management Agreement. Both the collateral assignment of the Management Agreement and the SNDA are immediately effective upon the loan closing. The result is that the Management Agreement is immediately subordinate to the rights of the Mortgagee, and this may give the Mortgagee certain rights immediately upon default, which is long before the foreclosure.

Because the SNDA and collateral assignment are immediately effective, it is important to understand what has been subordinated and what has not been subordinated to the Mortgagee. The strongest position for the Mortgagee is to have all the right, title and interest of Manager under the Management Agreement, *and* all rights of Manager relating to the use of funds in the various Hotel Operating Accounts subject to the SNDA. In this configuration, as soon as the Mortgagee begins to exercise its rights and remedies under the loan documents, it can insert itself into Manager's operation of the Hotel, including such matters as the disbursement of funds from FF&E Reserve accounts and even the use of Operating Accounts. Admittedly, a reasonable and experienced Mortgagee with an understanding of the hospitality industry is not likely to undermine Hotel operations that are in accordance with the Management Agreement and the approved Annual Budget, and result in the preservation of the value of the asset as a going concern, including its position in the market place and overall value. If the objective of the Mortgagee is to regain possession and control of the Hotel in order to sell it at foreclosure, any action that would diminish the value of the Hotel would be counter-intuitive, counter-productive and illogical. Nevertheless, it has been known to occur, and when Mortgagees insert themselves into Hotel operations post-default but pre-foreclosure, the job of Manager is that much more difficult.

Compare this to other situations where the SNDA either does not cover the Hotel's Operating Accounts or provides a series of carve-outs to permit Manager to continue to operate the Hotel in accordance with the terms and conditions of the Management Agreement. The Mortgagee is still present and interested, but the level of scrutiny of Manager is directed primarily to assessing conduct against the standards set forth in the Management Agreement, which most Mortgagees now underwrite and approve prior to closing in any event, instead of undertaking a de novo examination into the conduct of Manager. In the best of circumstances for Manager, the Mortgagee recognizes that there can be a time gap of many months, even years, between the acceleration of the indebtedness under the note and the foreclosure sale, and during that period of time,

Manager must continue to operate the Hotel in accordance with the Management Agreement. It is important to appreciate that all of this will occur within the context of then prevailing market conditions, and Manager, Owner, and Mortgagee may each have the stronger bargaining position from time to time.

The companion to the discussion regarding subordination is the discussion about non-disturbance. Remember the context of long-term Management Agreements and the value of management companies being essentially the value of their Hotel Management Agreements. This sets the foundation for the critical importance to Manager of gaining commitments from the Mortgagee that essentially say to Manager, "no matter what our borrower, the Hotel Owner, may do or fail to do under the loan documents, so long as you, Manager, are behaving as required under the Management Agreement, not only will we, the Mortgagee, leave you alone, but anyone who steps into the shoes of Owner through us will also leave you alone, and you, Manager, will be able to get the benefit of the bargain you thought you made with Owner." Some of the more common elements of non-disturbance are that:

- Manager will not be made a party to the Mortgagee's actions against Owner.
- The Management Agreement will not be terminated by the Mortgagee or a subsequent Owner acquiring the Hotel through foreclosure.
- Manager will be permitted to continue to operate the Hotel under the Management Agreement.

Another aspect of non-disturbance benefiting Manager is the so-called spring back provisions. If the Management Agreement was wrongfully terminated by Owner, or if Manager was precluded from operating the Hotel in a manner not contemplated by the Management Agreement, then Manager's right to manage the Hotel under the Management Agreement will "spring back" when a subsequent Owner acquires the Hotel through foreclosure. One way to think of it is the prevention of mischief by an Owner or Mortgagee to gain an unencumbered Hotel, free from the encumbrance of the Management Agreement, by wrongfully terminating the Management Agreement prior to foreclosure. The subsequent Owner still acquires the Hotel encumbered by the wrongfully terminated Management Agreement, through the obligation to enter into a new Management Agreement on the same terms as the prior Management Agreement for the remainder of the term of the prior Management Agreement.

Most SNDA forms are generally fair in that the subsequent Owner will avoid liability for most of the failures of its predecessor. These include the prior Owner's defaults under the Management Agreement, with the understanding that there are certain omissions or failures by a prior Owner that are likely ongoing and will need to be remedied by the subsequent Owner as a matter of continuing Hotel operations and good business. For example, if the prior Owner failed in its obligations to provide cash to cover a shortfall, the subsequent Owner will not be directly liable to Manager if Manager advanced funds, but the subsequent Owner would inherit the responsibility to insure that all reserves are fully funded and that all operating accounts are at proper levels.

Properly dealing with the SNDA can be very complex and time consuming. The relationship and interplay between the loan documents and the Management Agreement require a solid understanding of both in order to assess and address the wide variety of potential concerns. In addition, the SNDA can be, and often is, used as the vehicle for the modification of the Management Agreement either temporarily, for the term of the loan as affecting the current Mortgagee only, or permanently, for the term of the Management Agreement, as a form of Management Agreement amendment. With so much to do in the SNDA it remains one of the great mysteries of life in the law that the SNDA is relegated to the end of the deal and rises up like a great wave on an endless sea after the loan documents are fully negotiated and as the parties believe they are on the last road to the closing table.

The final word on the SNDA is about developing best practices. One important element of best practices is to place the development, structuring, form fights, and negotiation of the SNDA more at the core of the entire transaction rather than at the tail end.

For example, a recent development has been the creation of a revised form of SNDA that is structured as a two party instrument between Manager and Mortgagee. Owner joins in the SNDA as a joinder party to memorialize Owner's consent to the terms and conditions agreed to by Manager and Mortgagee. This is very logical because Manager and Mortgagee are the two parties with the most at stake after Owner has defaulted on its loan. Mortgagee is likely to appoint a receiver for the Hotel, so these two parties in interest must establish a methodology for the continued operation of the Hotel while Mortgagee implements its exit strategy. Regardless of how the SNDA is approached, the likelihood is that the process will be timelier and the product will be better if this part of the Management Agreement negotiation is commenced earlier in the process.

The SNDA in detail

Most SNDA forms include as a preliminary matter Owner's grant and transfer to Mortgagee of Owner's rights with respect to the key components of the relationship with Manager, such as the following:

- Owner's right, title, and interest in the Management Agreement.
- Owner's rights to proceeds under the Management Agreement.
- Owner's rights to the accounts that either Owner or Manager maintains under the Management Agreement.

Although Owner's assignment included within the SNDA is intended to operate as an absolute and immediately effective assignment and not merely the granting of a security interest to Mortgagee, until the occurrence of an uncured Event of Default by Owner under the Loan Documents, Owner retains the benefits of the Management Agreement and the other interests identified above that were assigned to Mortgagee. If an Event of Default occurs and is not timely cured, the rights granted to Owner will be either automatically revoked or revocable in Mortgagee's sole election, and this frees Mortgagee to exercise any and all rights and remedies under the Loan Documents and the SNDA. Unless and until Mortgagee exercises its rights under the SNDA, Mortgagee has no obligations with respect to the Management Agreement, and then only to the extent the SNDA exposes Mortgagee to Management Agreement obligations.

SNDA forms will be negotiated on a case-by-case basis with respect to exactly what is being subordinated. Mortgagees often express the requirement that Manager's interests in and to the Hotel under the Management Agreement, including the right to receive payment of fees are subject and subordinate to the lien of the Mortgage. Managers will usually agree that the Management Agreement does not grant Manager an interest in the real estate and therefore the lien of the Mortgage is senior to the Management Agreement, but it does not necessarily follow that Manager will agree that its right to receive payment of the fees under the Management Agreement is junior to the Mortgagee's right to be paid its debt service payments. This negotiation often results in any Incentive Management Fee being subordinated to Mortgagee's rights, but not the Base Fee and the use of Operating Accounts and FF&E Reserves as provided for under the Management Agreement and the approved Annual Budget. Despite Owner's Event of Default, there is still an operating business to be managed, and Manager is responsible for discharging the management function, for which Manager should be compensated by at least the Base Fee.

Mortgagees will agree not to name Manager as a defendant in the Foreclosure, unless joining Manager is necessary to foreclose the lien of the Mortgage, or otherwise act in a manner inconsistent with the provisions of the SNDA with respect to elements of the Management Agreement that are not subordinated to the Mortgage.

When an SNDA contains non-disturbance provisions for the benefit of Manager, Mortgagee or any Subsequent Owner acquiring title to, or assuming possession or control of, the Hotel, agrees (and Mortgagee's agreement is stated to be binding on all Subsequent Owners) to:

- recognize Manager's rights under the Management Agreement;
- not disturb Manager's right to manage and operate the Hotel under the Management Agreement; and
- assume all of the obligations of the "Owner" under the Management Agreement that are continuing or that arise after the Foreclosure Date, or the date of acquisition of title to the Hotel, whichever occurs later, pursuant to a written assumption agreement reasonably acceptable to Manager.

The funding of monies Manager forecasts to be needed upon and after the Foreclosure can be a challenge. One way to address this is for Mortgagee, should it become the Subsequent Owner, to have the option, rather than the obligation, to fund the amounts Manager projects will be required, and if Subsequent Owner elects not to fund those amounts, Manager can terminate the Management Agreement. The obvious Manager concern then is that the Subsequent Owner might intentionally withhold funds to drive Manager out of the Hotel.

Recent events caused many participants in Hotel transactions to pay much closer attention to receiverships. Some SNDA forms failed to address court appointed receivers or left vague whether the receivership was an initial Foreclosure followed by a second Foreclosure when the Hotel was later sold to a new Owner. SNDA forms should address receiverships, and when they do, the elements often present in the SNDA include the following:

- Mortgagee's covenant to provide prior notice to Manager if Mortgagee intends to seek appointment by a court of a receiver to assume possession or control of the Hotel, including detailed information identifying the name and location of the court, together with the proposed order appointing the Receiver. The intent is for Manager to be able to confirm that the order is consistent with the Management Agreement.
- During the pendency of the Receivership Mortgagee agrees not to direct the Receiver to violate Manager's rights under the Management Agreement, particularly if the SNDA contains non-disturbance of Manager.

These Mortgagee covenants as to the Receiver are consistent with an SNDA with non-disturbance because Mortgagee would have, through the SNDA, recognized Manager's rights under the Management Agreement, giving Manager the continuing right and authority to operate the Hotel pursuant to the Management Agreement. Mortgagees should forebear from exercising the role of the Hotel Owner from the posture of being the Mortgagee for a variety of reasons, including potential Mortgagee liability claims, and let the Receivership run its course.

This is less challenging to Manager if non-disturbance is included in the SNDA. If there is no non-disturbance and Manager is at risk of termination, the entire dynamic of the relationship of the Mortgagee and Manager is open for evaluation and negotiation.

SNDA forms that afford Manager non-disturbance will come with a price to Manager of being obligated to attorn to Mortgagee or a Subsequent Owner. Once Subsequent Owner acquires title to the Hotel, or assumes or obtains possession or control of, the Hotel, Manager will attorn to the Subsequent Owner. Manager will then remain bound by all of the terms, covenants, and conditions of the Management Agreement, but, of course, also enjoy all rights afforded to Manager under the Management Agreement, for the balance of the remaining term, including any renewal terms, with the same force and effect as if the Subsequent Owner were the "Owner" under the Management Agreement at the time originally executed.

The SNDA should provide for the possibility that the Subsequent Owner is not a permitted transferee or not willing to assume the Management Agreement. Mortgagee's concern is to have an exit strategy from a loan gone bad, so Mortgagee will want the ability to sell the Hotel to a Subsequent Owner under almost any circumstances. Manager has a different focus and while wanting to continue to manage the Hotel pursuant to the Management Agreement, must be cognizant of its new Owner. This is usually resolved by Manager's termination right. The result is that, if at the time a Subsequent Owner acquires title to, or assumes or obtains possession or control of, the Hotel the Subsequent Owner would not qualify as a permitted transferee under the applicable section of the Management Agreement, or the Subsequent Owner will not assume the Management Agreement, Manager will not be obligated to attorn and Manager will have the right to terminate the Management Agreement, subject to a relatively short period of time to exercise that termination right.

To protect the collateral security for its loan, Mortgagee will build into the SNDA a Manager obligation to deliver notice of Owner defaults under the Management Agreement and the right to cure those defaults. Upon any default by Owner under the Management Agreement leading to Manager giving a notice of default to Owner, Manager will agree to provide simultaneous notice to Mortgagee. This notice might be the sole notice to Mortgagee or a first notice to Mortgagee to be followed by a second notice later. For example, Manager provides a first notice to Mortgagee when it provides notice to Owner. If Owner fails to cure the default within the applicable cure period under the Management Agreement, and Manager intends to terminate the Management Agreement, Manager would then send Mortgagee a second notice, stating Manager's intention to terminate the Management Agreement, and Mortgagee would have a further right to cure Owner's Default for an extended period of time. Once Owner fails to cure the default in a timely fashion, Manager can proceed to exercise its rights and remedies under the Management Agreement, including termination, but those actions would be subject to Mortgagee's cure right and Manager would have to stop its actions, and the effort to terminate the Management Agreement would be rendered null and void if Mortgagee were to effect the cure during its extended cure period.

Many SNDA forms specifically provide that for a notice given by Manager to Owner to be effective as a notice under the Management Agreement a duplicate notice to Mortgagee must be provided. Any cure or performance by Mortgagee must be accepted by Manager as if it were performed by Owner.

Similarly, if Manager received a notice from Mortgagee that an Event of Default has occurred under the Mortgage, permitting Mortgagee to act for Owner, Manager will comply with Mortgagee's exercise of Owner's rights under the Management Agreement. By virtue of the language of most SNDA forms, and without the necessity of a separate power of attorney, Mortgagee has the full right and power in accordance with the Loan Documents to enforce directly against Manager all obligations of Manager under the Management Agreement and to essentially act as Owner. When Mortgagee acts in this manner and expends money to do so, all proceeds spent by Mortgagee represent additional advances under the Loan Documents. In the hospitality area, Mortgagees are often confronted with painful decisions that require Mortgagee to advance its cash as a protective advance on Owner's behalf. These advances by Mortgagee are generally demand obligations of Owner under the Loan Documents.

The notice and cure provisions already described for the benefit of Mortgagee should be balanced with similar provisions to benefit Manager. If Mortgagee sends Owner a notice of default under the Mortgage, then Mortgagee will be obligated to simultaneously send Manager a copy of the notice and Manager will have a period of time to take action, but usually not a cure right. Manager might, through specific negotiations, obtain a right to cure Owner's loan defaults or acquire by purchase Mortgagee's interest in the loan for an amount equal to the outstanding principal amount plus any accrued but unpaid interest.

Well-crafted SNDA forms recognize that Mortgagee might accelerate the repayment of the loan in the face of an Owner default under the Loan Documents, but not want to terminate the Management Agreement. At that point Mortgagee has a significant concern over the ongoing capital needs of the Hotel, but also has to respect the non-disturbance of Manager. In this more interesting context, Manager will require continued control over the Operating Accounts and the FF&E Reserve pursuant to the Management Agreement and the Annual Budget. Many Managers will, within a short period of time after Manager receives notice that Mortgagee has accelerated the debt, provide Mortgagee with an analysis of the working capital position of the Hotel for the balance of the Fiscal Year and some period of time thereafter, often 12 months, together with the amount of working capital that the Hotel will need for the next three months. While Mortgagee will want distributable cash to be paid to Mortgagee to pay Mortgagee's enforcement costs and reduce the unpaid balance of the loan, Manager will make that type of distribution only after providing for the anticipated near term working capital and reserve needs of the Hotel.

The nature of the SNDA as a collateral assignment carries meaning that goes deeper than the mere assignment by Owner to Mortgagee. Manager covenants in the SNDA to pay to Mortgagee distributions under the Management Agreement that would otherwise be payable to Owner if Manager receives notice from Mortgagee to do so. Manager is free to comply with Mortgagee's notice without investigation and without confirming whether Owner is in fact in default under the terms of the Mortgage.

Manager and Mortgagee may each need an estoppel certificate from the other during the term of the loan. To accommodate this, most SNDA forms include mutual provisions to provide an Estoppel Certificate. For example, Mortgagee may need a Manager's estoppel certificate to the effect that:

- the Management Agreement is unmodified and in full force and effect (or if there have been modifications, that the same, as modified, is in full force and effect and stating the modifications);
- management fees due and payable under the Management Agreement have been paid through a stated date; and
- to the best knowledge of Manager:
    - there is or is not a continuing default by Owner in the performance or observance of any covenant, agreement, or condition contained in the Management Agreement; and
    - there shall or shall not have occurred any event that, with the giving of notice or passage of time or both, would become a default, and, if so, specifying each default or occurrence of which Manager has actual knowledge.

Manager may need a Mortgagee's estoppel certificate to the effect that:

- to the best knowledge of Mortgagee, Mortgagee's consent has not been requested with respect to any modification of the Management Agreement; and
- to the best knowledge of Mortgagee, Mortgagee has or has not received any notice of default regarding the performance by Manager or Owner of their respective performance or observance of any covenant, agreement, or condition contained in the Management Agreement.

Although the provisions of the SNDA will be fully effective and binding between the parties and any Subsequent Owner without the execution of any further instruments by any party, each party to the SNDA should have a continuing right to request any other party to execute documentation (in form reasonably satisfactory to all signing parties) confirming the provisions of the SNDA.

Dealing with the proceeds from casualty insurance or condemnation proceedings is often a material aspect of arriving at an agreed upon SNDA. The dividing line is rather simple. Mortgagee wants the parties to agree that any insurance proceeds and condemnation awards received by Mortgagee with respect to the Hotel will be applied in accordance with the requirements of the Loan Documents, and Manager wants the application of proceeds to be in accordance with the Management Agreement. This is a case-by-case negotiation and the result can take many forms in the spirit of negotiation and compromise.

Comfort Letters

In the Franchise context, the relationship among Franchisor, Franchisee and Franchisee's Mortgagee is memorialized in the Comfort Letter, but Comfort Letters cannot generally be approached in the same manner as an SNDA.

Although each Franchisor will have its own form of Comfort Letter, the Comfort Letter generally begins with a statement that Franchisee and Mortgagee have requested

Franchisor to enter into the Comfort Letter. The preliminary covenants in most Comfort Letters include that:

- Franchisor will use reasonable, commercially reasonable or best efforts to give Mortgagee prior written notice, usually 30 days, of any voluntary surrender of the Franchise.
- Franchisor will use reasonable, commercially reasonable or best efforts to furnish Mortgagee with copies of default notices sent by Franchisor to Franchisee.
- Franchisor will allow Mortgagee a cure right, often 30 days, to cure Franchisee's default(s).

Of course, a Hotel is an operating business occupied by members of the public on a continuous basis, so some of the time periods in these Franchisor covenants will be abbreviated in the event of a health or life safety default. There is also a recognition that Mortgagee cannot cure Franchisee's bankruptcy, assignment for the benefit of creditors, or appointment of a receiver or trustee for Franchisee. In these situations, Franchisor is unlikely to act to immediately terminate the Franchise Agreement or preclude Mortgagee from acting to continue the effectiveness of the Franchise Agreement despite Franchisee's default.

The next crucial question is: what happens when Mortgagee acts to exercise its rights and remedies under the loan documents and protect its collateral security for the loan, meaning the Hotel? From the Mortgagee's perspective, Mortgagee wants the option to continue the Franchise Agreement and continue to operate the Hotel with the benefit of the Brand's systems and procedures. This may or may not be consistent with how Franchisor would like to proceed, particularly if the Hotel in question was not performing well or had an Area of Protection that interfered with Franchisor's plans for a newer Hotel or other Brand expansion. This is a case by case negotiation that may depend heavily upon where the Hotel is located. Franchisors may desire to keep all options open as to Hotels in more urban and densely populated areas with significant barriers to entry as compared to suburban locations.

If Mortgagee has the option to terminate the Franchise Agreement in the Comfort Letter, it will be exercisable within a specified period of time. If Mortgagee has that option and determines not to terminate the Franchise Agreement, Mortgagee will normally sign an assignment and assumption agreement in a form that Franchisor reasonably specifies, or a new Franchise Agreement in Franchisor's then applicable form, or an agreement whereby Franchisor will operate the Hotel, thus converting the franchise relationships into a Management Agreement, for a term equal to Franchisee's remaining term. The important point here is that Mortgagee and Franchisor will negotiate whether all of this is at Franchisor's option or Mortgagee's option. The other element of this that can catch a Mortgagee unaware is that the new Franchise Agreement will generally be in accordance with Franchisor's then prevailing standards, rates, requirements, and terms. This will need to be clearly understood by Mortgagee. Mortgagee will also want to pursue provisions that acknowledge that Mortgagee has become an involuntary Owner of the Hotel and for that reason would be exempt from paying an Application Fee for the new Franchise Agreement, and will not be required to perform a new PIP, renovation or upgrade of the Hotel, subject to curing existing quality deficiencies and continuing liability for renovations or upgrades required of other Franchisees in the Franchise system. Mortgagee should also anticipate curing then existing defaults under the Franchise Agreement that would represent a continuing default of Mortgagee the moment it becomes Franchisee, such as adequate funding of an FF&E Reserve or having the proper computer software for Franchisor's systems.

If Mortgagee had the option to terminate the Franchise Agreement, and determined not to exercise it, Mortgagee would have no obligations under the Comfort Mortgagee or the Franchise Agreement, and Franchisor would remove the Hotel from its system and demand the de-identification of the property.

If Mortgagee intends to acquire the Hotel and continue the Franchise Agreement until Mortgagee can sell the Hotel, Mortgagee might be wise to negotiate relief from the liquidated damages provisions of the Franchise Agreement. This would be a personal right for the benefit of Mortgagee to facilitate an anticipated future sale of the Hotel.

Mortgagee will make a number of covenants in favor of Franchisor under every Franchisor's form of Comfort Letter. The most common as are follows:

- To provide advance notice of any action by Mortgagee to:
  - commence foreclosure proceedings;
  - seek the appointment of a receiver;
  - seek an order for relief or take any action under federal or state bankruptcy laws;
  - accept a deed in lieu of foreclosure; or
  - take ownership or possession of the Hotel in any manner.

- To notify Franchisor if Mortgagee receives notice that another party has commenced foreclosure proceedings or sought the appointment of a receiver or commenced a petition for relief under state or federal bankruptcy laws.

Mortgagees are usually not qualified professional Hotel Managers. For that reason, if Mortgagee acquires the Hotel, Mortgagee will agree to engage a professional Manager or management company to operate the Hotel under the Franchise Agreement. To protect themselves, Franchisors will specifically provide in the Comfort Letter that the management company must be approved by Franchisor, and that the general Manager and other senior management personnel must complete or have recently completed Franchisor's training requirements. Many Franchisors maintain and make available to Mortgagees a list of approved management companies.

Consistent with the personal nature of the Franchise Agreement, Comfort Letters are assignable in only limited situations. Mortgagee will generally be able to assign the Comfort Letter only to a subsequent holder of the loan without Franchisor's consent, subject to certain further requirements, such as that the successor to Mortgagee:

- is a commercial bank, investment bank, pension fund, finance company, insurance company, trustee in a securitization or other financial company, or other financial institution or another type of established organization, so long as that established organization is not a Brand Owner under the Franchise Agreement or does not exclusively lend to a Brand Owner, primarily engaged in the business of making or holding loans and any fund or trust managed or serviced by any of the foregoing; and
- does not own, directly or indirectly, any equity interest in Franchisee or its constituent Owners.

Some Comfort Letters require the payment to Franchisor of a processing charge upon the sale or transfer of the Loan, together with the execution of an assignment and assumption agreement.

If a third party becomes Owner of the Hotel and the Franchise Agreement has not been terminated, that third party must make application for a new Franchise Agreement, including payment of the Application Fee, unless another legal requirement, such as a bankruptcy court plan or order provides to the contrary, such as relieving the new owner of any obligation to accept the Franchise Agreement. For example:

> Franchisee will acknowledge that the Comfort Letter was provided to Mortgagee at its request, and Franchisee will release Mortgagee and Franchisor of and from any and all actions, causes of action, suits, claims, demands, contingencies, debts, accounts and judgments whatsoever, at law or in equity, whether known or unknown, arising from Mortgagee's exercise of any of the rights granted under the Comfort Letter.

When the negotiation is completed, and a Comfort Letter executed and delivered, it will often provide Mortgagee with the ability to continue to operate the Hotel pursuant to the Franchise Agreement, or decline to do so and let Franchisor terminate the Franchise Agreement and remove the Hotel from Franchisor's system, essentially, Franchisor removes the flag or Brand from the Hotel, disconnects the Hotel from the reservation systems, demands the Hotel be de-identified, and then the Hotel must operate either as an independent Hotel or as part of another system. The situation described here is not immutable and is undergoing a change in some situations. Some Franchisors are beginning to view the Comfort Letter in the same manner as they view an SNDA. This is particularly the case when the Hotel is in a desirable location that Franchisor wants to retain despite the failure of Franchisee to succeed. This will be evidenced in a Comfort Letter that requires Mortgagee to continue the Franchise Agreement either through an assignment of the existing Franchise Agreement or the execution of a new Franchise Agreement with Mortgagee or its designee or successor. Treating a Franchisor's Comfort Letter like a Manager's SNDA will present challenges because the two relationships have materially different components, and use of a Franchise instead of management is often considered because the Liquidated Damages are set forth in the Franchise Agreement and represent a known and liquidated sum for termination of a Franchise Agreement, unlike a long-term Management Agreement with either no provisions for agreed upon Liquidated Damages or a significant fee for termination prior to the natural term of the Management Agreement. A Comfort Letter that is construed as an SNDA may be no comfort for Mortgagees, and they may decide to proceed without it and deal with Franchisor if and when the need arises.

7   The Technical Services Agreement

The Technical Services Agreement or "TSA" can be incorporated into the body of the Hotel Management Agreement, or can be a separate document through which Manager will deliver its Brand supervision services during the design plan preparation and construction or refurbishment stage. This agreement often includes pre-opening services, but incorporating both technical services and pre-opening services into the Hotel Management Agreement is not unusual.

Manager will use the TSA as a means of reviewing and approving all of the design plans for the Hotel and, to a lesser extent, the construction plans, to assure compliance with the Manager's design criteria. Owner, in turn, will use the TSA to manage the construction of the Hotel within an agreed upon construction budget, while ensuring that upon substantial completion, the Hotel will satisfy all of the requirements of the Brand. The Hotel, as an operating business, must be ready to open to the public before the Opening Date, and pre-opening services will address those functions and the respective rights and obligations of the Manager and Owner.

The general structure of an agreement that includes both technical services and pre-opening services will be explored in this chapter.

Some hotel management transactions will include a significant number of documents, many of which need not be executed when Owner and Manager begin their business relationship. However, it is common for the TSA and the HMA to be executed first and contemporaneously. Work under the TSA may commence immediately, while the HMA will control the relationship on and after the Opening Date of the Hotel.

Be careful for arrangements where the Manager under the Management Agreement and the Consultant under the TSA are affiliated companies but not the same company. Many Managers deliver technical and pre-opening services through an Affiliate. This will be relevant when the parties consider the identity of a creditworthy entity to stand behind certain obligations of the Consultant, such as indemnification of Owner, as well as the potential cross-termination of the TSA and the HMA.

Retention of services

Owner and Consultant should specify the Technical Services and Pre-Opening Services to be provided. This can be accomplished through recitation in the body of the TSA or through the use of attached schedules and exhibits creating a laundry list of services. The list can be further refined to indicate mandatory services that Consultant will provide and optional services that the Consultant will provide upon Owner's request, and, potentially, with a separate fee structure. Even though Consultant will be engaged to provide various services, it is Owner that is ultimately responsible for the development and construction of the Project, including permitting, design, budgeting, construction, scheduling, accounting and administration, and coordination between Consultant and any designer or contractor engaged by Owner, and, generally, the technical services provider will expect to be indemnified for any claims arising from Owner's failure to discharge these responsibilities.

Owner is responsible for constructing, furnishing and equipping the Project at its own expense and in accordance with a schedule that Owner and Consultant will mutually approve. When the Hotel is Branded and subject to Manager's Brand Standards, Owner's development of the Project will also be:

- to a level no less than the Brand Standard;
- in accordance with design documents that will have been reviewed and possibly approved by Consultant;
- in conformity with all applicable laws.

A representative example of Pre-Opening and Technical Services might include each of the following.

Pre-Opening Human Resources

These might include:

- identifying, appoint, assign, instruct, train and supervise a general manager and other personnel at the Hotel, but as contemplated by the Pre-Opening Budget;
- hiring a general manager and director of sales well in in advance of the Opening Date, within a range of compensation reflected in the Pre-Opening Budget.

Pre-Opening Marketing

This could involve assisting in the development of a marketing plan for the promotion and marketing of the Hotel, including a promotional plan for the sale of rooms and food and beverage services, public relations and communications activities, implementation of market research, sales program, reservations program and shared advertising program, and representation of the Project by international sales and corporate marketing personnel of Consultant and its affiliates.

Food and Beverage Operations

This might include assistance with respect to the implementation of the initial Food and Beverages Operations plan, including service concepts, decor, menus, and potential operators.

Interior Design Consultation/Review Services

This entails review and provision of comments to Owner on interior design concepts as developed by Owner's interior designer, and review and provision of comments to Owner on interior design documents and materials developed by the interior designer.

Construction Services

This includes such things as:

- assisting Owner with oversight of construction and/or build-out matters, such as, HVAC, fire safety systems, electrical systems, and communications systems;
- assisting Owner in the evaluation and approval of proposed material substitutions as submitted for value engineering or schedule requirements;
- review of model room design and construction documentation prior to commencement of model room construction;
- review and approve the model room on completion.

Licenses and Permits

This would involve assistance in the application for and procurement of all governmental or other approvals.

Initial Supplies

That is, providing Owner with the specifications and quantities for the initial OS&E, inventory, and operating equipment.

Computer Services

This might include data processing, telecommunications, office automation, and computer services for the Hotel.

Opening date

The TSA should specify a procedure for the determination of the date that the Hotel will begin accepting paying guests for rooms and food and beverages. This date could be unilaterally determined by Owner, or by Owner but with the reasonable consent of Consultant. In either situation, each of the following events generally must have occurred to achieve the Opening Date:

- substantial completion of the Hotel;
- issuance of a temporary certificate of occupancy or its equivalent;
- issuance of a certificate from the architect certifying that the Hotel has been substantially completed in accordance with the design documents;
- all FF&E and OS&E has been substantially installed at the Hotel;
- all required mechanical and electrical systems have been tested, commissioned and are functioning properly;
- all licenses and permits required for the operation of the Hotel have been obtained;
- adequate Working Capital has been furnished by Owner in accordance with the Management Agreement;
- the Hotel has been accepted by Consultant as being in a condition that will permit Consultant to operate the Hotel in accordance with the Brand Standard.

The final requirement must be carefully reviewed by Owner and Manager. In the context of a Hotel with a Brand, any Brand, the Hotel must have been completed in accordance with the Brand Standard. To achieve a mutually satisfactory result and avoid surprises as the Opening Date approaches, the best practice is for Owners and Managers to cooperate throughout the term of the TSA and the design and development process.

Certainly, a complete opening on the Opening Date with all rooms and venues operational is the goal. Nevertheless, the TSA can provide flexibility for the Hotel to partially operate prior to the time that all of the conditions have been satisfied if Owner and Manager can agree and the delivery of guest services would not be compromised.

Purchasing agent

Procurement of FF&E and OS&E initially occurs under the TSA. Owner may engage a third-party Procurement Agent or use the services of an agent recommended by or affiliated with the Consultant.

Personnel

When the Consultant is performing under the TSA, it may use its home office employees, both on and offsite as appropriate. The Consultant is responsible for the supervision of these employees, as well as their compensation. The one exception to this will be travel, housing and meal costs of those employees incurred in providing the services under the TSA, and these will be subject to reimbursement by Owner.

Pre-Opening Services

Pre-Opening Services will commence prior to the Opening Date and the period of time will be stated in the TSA. Depending on the size and complexity of the Hotel, the time period can be in a range of between six and 18 months prior to the anticipated Opening Date. Consultant should deliver the Pre-Opening Services in accordance with a Pre-Opening Budget that should include the following information:

- a preliminary budget and monthly expenditure chart for the Pre-Opening Period (as would be established in the TSA), including all pre-opening expenses, the expenses for the Pre-Opening Marketing Plan (this would also be established in the TSA), and all other expenses contemplated in the Pre-Opening Budget
- an organization chart of job positions for all pre-opening personnel and an on-boarding schedule
- a preliminary cost estimate and outline (including an outline of the timing of the implementation) of the critical elements of the Pre-Opening Marketing Plan
- a preliminary insurance coverage plan and cost estimate for Consultant's insurance program.

Comments to Pre-Opening Budget

The TSA will establish a procedure for Owner's approval of the Pre-Opening Budget. The most favorable situation is when Owner and Consultant cooperate on the development of the Pre-Opening Budget and reach mutual agreement on its contents.

Owner and Consultant should each act reasonably and work together to reach agreement on the Pre-Opening Budget as soon as reasonably possible. Consultant can be expected to require that Owner not unreasonably withhold or delay approval of any matter in the Pre-Opening Budget that is required for the Hotel to comply with the Brand Standard.

Bank account

The Consultant will be authorized to establish a segregated bank account on its behalf at a bank usually specified by Owner or Mortgagee, and subject to Mortgagee's requirements. Owner then deposits into the bank account the amounts specified in the Management Agreement and/or the TSA for pre-opening and pursuant to the Pre-Opening Budget.

Compensation

Consultant's compensation under the TSA is separate and apart from Manager's compensation under the Hotel Management Agreement. The compensation is a negotiated subject and can be formulated in a number of ways. Some Consultants use a flat fee, while others use a dollar amount per Key. The payment of the compensation is also negotiated. Payments can be monthly or include an initial payment, followed by a monthly payment and then concluded with a final lump sum payment of the balance upon substantial completion of the Hotel. The TSA should provide for the possibility that the Technical Services Fee would have been fully paid to the Consultant, but the project not be complete. In that case, and if negotiated as part of the TSA, Owner would be obligated to reimburse Consultant for all reasonable costs and expenses incurred by Consultant in connection with the project, together with further monthly installments of the Technical Services Fee until completion of the project. The TSA may also specify that if the components, scope or size of the project have materially changed in any material respect, then Owner would pay an increased Technical Services Fee in an amount either agreed upon or to be agreed upon by the Parties, based on the nature and magnitude of the change, extension or delay. The TSA can provide that if Owner and consultant reach an impasse on this payment, that the matter would be submitted to the Expert.

Reimbursement

Consultant's Technical Services Fee does not usually include reimbursable expenses. The TSA should obligate Owner to pay or reimburse Consultant for all reasonable, documented, and verifiable third-party costs and expenses incurred in connection with the performance of the Technical Services and Pre-Opening Services. The TSA should set out a reimbursement procedure for Consultant to follow, including a formal request for reimbursement that includes reasonable documentation of the request and a certification by Consultant to Owner that the costs have been reasonably incurred for proper execution of, and directly related to the performance of, the Technical Services or other services described in the TSA. Payment is usually made once each month within a stated period of time after Owner's receipt of the request, such as 30 days after receipt.

Indemnity

The TSA will include indemnification provisions very similar to those found in the Hotel Management Agreement. Owner will indemnify and hold Consultant harmless from claims, demands, damages, liabilities, costs, losses, penalties, fines and expenses (including reasonable attorneys' fees), often described as "Claims", arising out of or in connection with the Consultant's work under the TSA, except to the extent Claims were caused by the gross negligence or knowingly willful misconduct of Consultant's corporate personnel who perform the Technical Services onsite. The Consultant will indemnify and hold Owner harmless from and against all Claims arising out of Consultant's Gross Negligence or Willful Misconduct. Imputed liability under a TSA will often be limited, and this should be an element of the TSA that both Consultant and Owner discuss in the negotiation of the document. The TSA should specify whose acts and omissions will or will not be imputed to Consultant with respect to the TSA.

Termination

A non-defaulting party will seek the ability to terminate the TSA and seek other remedies in the event of a default on the part of either party. "Default" under a TSA often includes:

- the failure of a party to timely make any payment to the other that continues for some brief period of time, usually 10 days, after written notice;
- the failure of a party to perform under or to comply with any material provisions of the TSA that continues for a period of time, usually 30 days after written notice;
- various creditors' rights and bankruptcy matters such as making an assignment for the benefit of creditors, filing a voluntary bankruptcy petition, and being the subject of an involuntary petition alleging an act of bankruptcy or insolvency or seeking reorganization that is not finally dismissed a short period of time after filing.

The TSA should also address other matters that might be better suited to expedited arbitration, such as disputes with respect to the Pre-Opening Budget and/or amounts due under the TSA or with respect to Consultant's consent or approval rights.

Most TSAs have a series of milestones that lead up to the Opening Date and then a required Outside Opening Date or Outside Commencement Date. If the Opening Date fails to be achieved by the Outside Commencement Date, Consultant may have the right to terminate the TSA. This may trigger a termination of the HMA through the operation of cross-default or cross-termination provisions in both the TSA and the HMA.

Limitation on recourse and damage claims against consultant

Consultants are very interested in establishing limitations on the aggregate liability of the Consultant for any Claims under the TSA, other than those for fraud, gross negligence or willful misconduct. One of the more common formulations of this limitation is to provide that Consultant's damages limitation is the amount of any fees or other amounts reimbursed and paid to Consultant pursuant to the TSA. As implied earlier, this will not apply to any monetary claims, damages, costs, or expenses which are proven by final judgment to have resulted from the fraud, willful misconduct or gross negligence of Consultant.

Scope of consultant's advice

Consultants are often careful to specify that when Consultant reviews something and offers input or advice to Owner, that is for the limited purpose of indicating whether the project is constructed, furnished and equipped in compliance with the Brand Standard and whether the project will be acceptable to Consultant from a functional and aesthetic point of view. It is not a representation as to the adequacy, quality, safety or cost of the project or its construction, operation or maintenance, and it would not create a duty on the part of Consultant that could give rise to any cause of action against Consultant by Owner or any third party based on any alleged deficiency in the adequacy, quality, safety, or cost of the project or its construction, operation or maintenance.

Limitation on recourse and damage claims against owner

Consultant is required to look solely and only to Owner's interest in the project for the payment of any amount to be paid by Owner.

Assignment

TSAs are generally not assignable without the prior written consent from the other Party other than to an Affiliate and concurrently with an assignment of the Management Agreement, to an assignee or transferee permitted under the Management Agreement.

Glossary of terms

AAA   *see* American Arbitration Association

Absentee ownership   an option offered by some Franchisors that allows a Person to own a Franchise without being actively involved in its day-to-day operations.

Accounting period   specific period of time during the Term, such as each calendar month, in which accounting statements are provided by Manager to Owner, such as Monthly Reports.

ADA   *see* Americans with Disabilities Act

Additional Insured   a Person or organization that enjoys the benefits of being insured under an insurance policy, in addition to whoever originally purchased the insurance policy. The term generally applies within liability insurance and property insurance, but is an element of other policies as well. Most often it applies where the original named insured provides insurance coverage to additional parties so that they enjoy protection from a new risk that arises out of the original named insured's conduct or operations. An additional insured often gains this status by means of an endorsement added to the policy that either identifies the additional Party by name or by a general description contained in a "blanket additional insured endorsement".

Adjusted Gross Operating Profit   Gross Operating Profit minus the Base Fee.

Adjusted Net Operating Income   amount determined in accordance with the Uniform System of Accounts for the Lodging Industry, on an accrual basis, excluding interest income and expense, depreciation and amortization, but after deducting the amount contributed to the annual FF&E Reserve and an agreed on Owner's Priority.

ADR   *See* Average Daily Rate

Affiliate   as to any Person, any other Person that, directly or indirectly, controls, is controlled by or is under common control with such Person. For purposes of this definition, the term "control" (including the terms "controlling," "controlled by," and "under common control with") of a Person means the possession, directly or indirectly, of the power: (i) to vote more than an agreed on percentage of the voting stock of such Person; or (ii) to direct or cause the direction of the management and policies of such Person, whether through the ownership of voting stock, by contract or otherwise.

AGOP   *see* Adjusted Gross Operating Profit

Agreement   Purchase Agreement or a Purchase and Sale Agreement.

American Arbitration Association   not-for-profit organization in the field of alternative dispute resolution, providing services to individuals and organizations that wish to resolve conflicts without litigation, and one of several arbitration organizations that administers arbitration proceedings.

Americans with Disabilities Act   a law enacted by the U.S. Congress in 1990, as amended.

Annual Budget   budget agreed on between Owner and Manager with projections for each Accounting Period of the forthcoming Fiscal Year.

Annual Financial Statement   statement, in reasonable detail, summarizing the operations of the Hotel for the immediately preceding Fiscal Year, often accompanied by a certificate of Manager's chief accounting officer certifying that the Annual Financial Statement is true and correct. Also known as Annual Operating Statement.

Annual Operating Statement   *see* Annual Financial Statement

AOP   *see* Area of Protection

Applicable Law   any federal, state or local law, code, rule, ordinance, regulation, or order of any governmental authority or agency having jurisdiction over the business or operation of the Hotel or the matters which are the subject of the Operative Agreement, including, without limitation, the following: (i) any building, zoning or use laws, ordinances, regulations or orders and (ii) all applicable Environmental Laws.

Application Fee   fee required to be paid by Franchisee to Franchisor upon application for a Franchise.

Area Developer   developer who agrees to open a certain number of Franchise units in a large territory within a specified time period. Such developer may open and operate the units themselves or recruit other Franchisees to open the units.

Area of Protection   specific area in which Licensor/Franchisor and its Affiliates are prohibited from granting a license for the Brand or promoting or marketing or publicly announcing any affiliation with any individual or entity in connection with the ownership, development, sale, marketing, advertising, promotion or operation of any other hotel project of the Brand. Also known as Radius Restriction or Exclusive Territories.

Average Daily Rate   refers to a statistical measure of hotel performance through determining the average rental income for each paid occupied room during any given time period.

Base Fee   base management fees earned by the Manager. The Base Fee is drawn by the Manager monthly and considered an Operating Expense of the Hotel. Also known as Base Management Fee.

Base Management Fee   *see* Base Fee

Beverage Management Agreement   agreement pursuant to which Manager or an Affiliate provides beverage management services at the Hotel.

Blocked Person   person, group, entity, or nation named by any Executive Order or the United States Treasury Department as a "Specially Designated National and Blocked Person," or other banned or blocked person, entity, nation, or transaction pursuant to any law, order, rule, or regulation that is enforced or administered by the Office of Foreign Assets Control.

Board   state and local agencies for the issuances of the Required Liquor Licenses.

Borrower   Party that borrowed funds from Mortgagee pursuant to the Loan Documents.

Boutique Hotel   intimate, usually luxurious, unique, or quirky and upscale hotel environment for a very particular clientele. Boutique Hotels are typically small with fewer than 100 Guest Rooms and often contain luxury facilities in unique or intimate settings with full service accommodations.

Brand　name, logos, know-how, intellectual property, and all form or manner of Brand Standards employed by a Branded Hotel Manager or Franchisor in the identification and operation of a hotel within the system of operations of the Branded Management Company. For example, Hilton is a recognized hotel Brand with a series of Brand Derivatives. (Its names and logos are included for information purposes at the companion website. Please visit www.routledge.com/cw/migdal to view the examples.)

Brand Derivatives　Hotel that is similar to the Brand that includes the Brand name as part of the name of the Hotel but is free from a Brand-specific Area of Protection or Radius Restriction.

Brand Marketing and National Sales Contribution　fair and equitable share of the costs for Brand Marketing Services and the National Sales Program to be paid by Owner to Licensor or its Affiliates.

Brand Marketing Services　affiliation of the Hotel with the chain of hotels operated under the System and the activities undertaken by Manager in connection therewith, including, by way of example, protection of the name and mark utilized in connection with the System, developing, preparing, producing, directing, administering, conducting, maintaining, and disseminating advertising, marketing, promotional, and public relations materials, programs and campaigns, conducting market research and similar activities.

Brand Owner　Licensor or the Franchisor, as applicable.

Brand Standards　any one or more of the following categories of standards: (i) operational standards (such as, guest services, quality of food and beverages, cleanliness, staffing and employee compensation and benefits, Centralized Services, frequent traveler programs and other similar programs, and any third-party managed or leased areas); (ii) physical standards (such as, overall quality of the Hotel, FF&E and Fixed Asset Supplies, frequency of FF&E replacements); and (iii) technology standards (such as, those relating to software, hardware, telecommunications, systems security and information technology); each of the standards is intended to be consistent with the applicable standard in effect at all other Branded Hotels.

Branded　operating under a Brand.

Branded Hotel　a Hotel operating under a Brand.

Branded Hotel Manager　operator or manager of a Branded Hotel. *See also* branded hotel operator; branded manager

Branded Hotel Operator　operator or manager of a Branded Hotel. *See also* Branded Hotel Manager; Branded Manager

Branded Identities　Hotel funds or Owners who become associated with a Brand through the acquisition of Hotels associated with a particular Brand.

Branded Management Company　management company which is affiliated with a Brand.

Branded Manager　*see* Branded Hotel Manager; Branded Hotel Operator.

Branded Residences　residential units managed by Branded hotel chains and are part of a Hotel compound. Branded residential realty is not a new idea. Since as early as the 1920s Hotels have offered private residences as well as transient guest rooms. Originally Branded Residences were residential properties linked to an adjacent Hotel. Since then the market has matured and widened and developers are now offering an incredible range of services. Branded residential real estate offers Buyers the opportunity to own property in a Branded Residence, enjoy the benefits of

Hotel services and amenities and permit the rental of the Residence as part of the room inventory.

Branded Residential Realty   *see* Branded Residences

Branding   associated with a Brand.

Building   building in which the Hotel is located.

Building Estimates   amount the Manager estimates Owner will have to expend to maintain and operate the Building.

Buyer   purchaser of the Hotel.

Capital budget   detailed schedule of the amounts to be deposited to the Reserve and all anticipated expenditures to be made from the Reserve or otherwise proposed to be funded by Owner for non-routine or major repairs, alterations, improvements, renewals and replacements to the Hotel, such as, the structure, the exterior façade, the mechanical, electrical, heating, ventilation, air conditioning, plumbing or vertical transportation elements of the Building, and other non-routine repairs and maintenance to the Building.

Capital Call   legal right of an entity to demand a portion of the money promised to it by its members or shareholders. Also known as drawdown or a capital commitment.

Capital Improvements   any (i) addition to or renovation of the Hotel that results in an increase in the number of Guest Rooms or meeting facilities or other facilities, (ii) reconfiguration of or renovations to the interior or exterior of the Hotel that are of a capital nature, (iii) improvements of a capital nature required to bring the Hotel into compliance with Legal Requirements, the physical aspects of the System Standards or any Mortgage, or (iv) improvements set forth in an approved Capital Budget.

Cash Collateral Account   bank account in the name of a Borrower that serves to secure and service a Loan. While cash and check deposits are made into this account, it is considered essentially a zero-balance account and it cannot be drawn upon like a checking account. With a cash collateral account, money is deposited in a lockbox account; when the funds are collected and the deposits have cleared, the debt served by the account is reduced or remaining cash can be returned to the Borrower.

Cash Management Agreement   agreement pursuant to a Borrower's Loan Documents to set forth the terms and conditions upon which the Gross Revenues will be deposited, maintained and withdrawn from a lockbox fund.

Category   classification of hotel for the purpose of differentiation within a Brand, such as airport hotel, convention hotel, or resort hotel.

Centralized Services   services provided by Manager on a centralized basis to all or substantially all of Manager's projects on a consistent basis at-cost with no mark-up (i.e., accounting, information technology, national sales and marketing, revenue management).

Centralized Services Agreement   agreement by and between Owner and Manager or an Affiliate that provides the terms and conditions pursuant to which Manager or an Affiliate will provide Centralized Services to the Hotel.

Centralized Services Fee   a fee that is often separated from the Marketing Fee and is sometimes set forth in a Centralized Services Agreement rather than incorporated into the Hotel Management Agreement. The Centralized Services Fee compensates the Manager for Centralized Services to the extent compensation is not already included in the Base Management Fee.

CFO *see* Chief Financial Officer

Chain Acquisition an exception to the Radius Restriction whereby an Owner may acquire a chain of Hotels of the same chain scale.

Chief Financial Officer a corporate officer primarily responsible for managing the financial risks of the entity. This officer is also responsible for financial planning and record-keeping, as well as financial reporting to higher management. In some sectors, the CFO is also responsible for analysis of data. The title is equivalent to finance director, a common title in the United Kingdom. The CFO supervises the finance unit and is the chief financial spokesperson for the organization. The CFO reports directly to the President/Chief Executive Officer (CEO) and directly assists the Chief Operating Officer (COO) on all strategic and tactical matters as they relate to budget management, cost benefit analysis, forecasting needs, and the securing of new funding. The CFO may be a member of the governing board of the entity.

Claims all losses, damages, charges, liabilities (direct or indirect), claims, expenses (including reasonable attorneys' fees and expenses) and suits or other causes of action of any nature whatsoever arising from or in any way connected with certain acts or omissions as specified in the Operative Agreement such as (i) the Hotel or the performance of Manager's obligations under and in accordance with the terms of the Operative Agreement, (ii) any other acts performed by Manager at the direction of Owner or pursuant to the authority granted to Manager under the Operative Agreement, (iii) the breach of any material provision of the Operative Agreement by Owner, (iv) Owner's failure (other than by reason of Manager's default under or breach of the Operative Agreement) or refusal to comply with or abide by any Applicable Laws, unless, following a final determination on the merits pursuant to the dispute resolution procedures prescribed in the Operative Agreement, it is determined that the Claim was attributable to Manager's fault, and (v) the operation of the Hotel, including Owner's maintenance of licenses and permits and compliance with Applicable Laws and Owner's assumption and performance of assignable contracts after any termination or expiration of the Operative Agreement.

Closing sale and purchase of the Hotel or Property pursuant to the Purchase and Sale Agreement.

Closing Date date on which Closing occurs.

Collateral Assignment of Management Agreement agreement whereby the Hotel Management Agreement is assigned to the Mortgagee as part of the Loan Documents.

Collective Bargaining Agreement any labor contracts, collective bargaining agreements or similar written agreements with labor organizations.

Comfort Letter a letter intended to govern the ability of a Mortgagee to operate a Hotel under a Brand name after a Borrower default or Foreclosure, receivership, and eventual sale.

Commercially reasonable efforts a term incapable of a precise definition and will vary depending on the context in which it is used. It is based on a standard of reasonableness, which is a subjective test of what a reasonable person would do in the individual circumstance, taking all factors into account. "Commercially reasonable efforts" refer to efforts that use a standard of reasonableness defined by what a similar person would do as judged by the standards of the applicable business community. The test for "commercially reasonable efforts" is less stringent than that imposed by the "best efforts" clauses contained in some agreements. A business may give reasonable consideration to its own interests, exercising discretion within its good faith business, judgment, in devising a strategy for achieving its ultimate goal. "Commercially reasonable efforts" has traditionally been consistent with good faith business judgments.

Company-Owned Units   those Franchised properties that are owned and run by the parent company (the Franchisor), rather than by Franchisees.

Competitive Set   agreed on specific set of hotels set forth in the Hotel Management Agreement, which are typically in the same market segment and geographic market as the Hotel, and intended to serve as benchmarks for the Hotel to measure its own performance. The Competitive Set may be modified, from time to time, so that it continues to represent a set of Hotels reasonably competitive with the Hotel in question.

Competitor   any Person who is engaged in the business of operating (as distinguished from directly or indirectly owning) hotels or resorts in competition with Manager or its Affiliates. A Person is not deemed to be in the business of operating hotels or resorts in competition with Manager or its Affiliate solely by virtue of holding a mortgage or mortgages secured by hotels or resorts.

Conference Revenues   revenues from conferences, receptions, meetings, and other functions occurring in any conference, banquet or meeting rooms in the Hotel, including usage charges and related taxes, food and beverage sales, valet parking charges, equipment rentals, and telecommunications charges.

Confidential Information   any confidential, proprietary or non-public information relating to the assets, trade secrets, methods, financial affairs, or business of the Party to whom such information is confidential, such as (i) non-public information of the Party or any Affiliates, (ii) non-public information provided by a Party or its Affiliates to the other Party; and (iii) the books and records (and non-public information contained therein).

Construction Management   management of the construction of the Hotel.

Construction Management Fee   *see* Development Fee

Construction Services   those services provided by Manager or an Affiliate for management of construction of or for the Hotel as frequently set forth in a separate Development Agreement or Construction Management Agreement.

Consultant   provider of services under a Technical Services Agreement.

Controllable Expenses   all Operating Expenses other than Uncontrollable Expenses.

Controlled Account Agreement   an agreement generally used to perfect a security interest on a depository account maintained at a bank, where the owner of the account becomes a debtor to a third party (Secured Party), and gives them authority to have disposition over the account. The owner of the account will have no say on the account, the Secured Party will provide instructions as to what to do with the funds, and the bank or depository financial institution controls the funds and acknowledges that they will only follow instructions of the Secured Party. The documents will be signed by the Secured Party, the depository entity (Bank), and the owner of the account (Debtor).

Controlling Interest   either (i) the right to exercise, directly or indirectly, more than an agreed on percentage of the voting rights attributable to the ownership interests in any Person (through ownership of membership interests, partnership interest, shares or by contract), or (ii) the possession, directly or indirectly, of the power to direct or cause the direction of the management or policies of any Person.

Conversion   option offered by some Franchisors to convert their existing independent business into a Franchise, or to convert a Management Agreement to a Franchise Agreement.

Corporate Personnel   those employees of Manager or its Affiliates (other than Hotel Employees) who are not ordinarily located at the Hotel, such as, home office and regional employees of Manager.

CPI Adjustment   when applied to any specific dollar amount means that dollar amount increased by a percentage equal to the amount, if any, by which the CPI Index in effect as of the end of the month preceding the date on which the CPI Adjustment is being computed exceeds the CPI Index in effect as of the end of the Fiscal Year in which the Opening Date falls.

CPI Index   Consumer Price Index for that area that is the smallest area encompassed by the Consumer Price Index, which also includes the area in which the Hotel is located published from time to time by the United States Bureau of Labor Statistics.

Cut-Off Time   agreed on time as set forth in the Purchase Agreement when the inventories of the Hotel are counted and values allocated to Seller and Purchaser accordingly.

Debt   outstanding principal amount set forth in, and evidenced by, any mortgage, together with all interest accrued and unpaid thereon and all other sums due and payable to a Mortgagee and any applicable mezzanine lender in respect of the loans made by Mortgagee and any applicable mezzanine lender to Owner from time to time.

Debt Service   cash that is required for a particular time period to cover the repayment of interest and principal on a debt. Debt Service is often calculated on a yearly basis. Companies may have outstanding loans or outstanding interest on bonds or the principal of maturing bonds that count towards the company's debt service. A company that is not able to make payments to service the Debt is said to be "unable to service its debt." In the context of an HMA "Debt Service" means, in each Fiscal Year, the Debt Service (for principal, interest, fees, and commissions) that (i) would be payable by Owner with respect to any bank facility or any other kind of borrowings for financing the construction of the Hotel and/or any refinancing thereof and/or (ii) would be provisioned pursuant to the bank facilities or other borrowings under the preceding clause (i).

Debt Service Coverage Ratio   amount of cash flow available to meet annual interest and principal payments on Debt.

Default   *see* Event of Default

Design Fee   fee for the provision of technical and design services by Manager. The Design Fee is typically a set fee paid monthly or quarterly during the development or redevelopment of a Hotel.

Development Agreement   agreement whereby the Hotel Owner engages the Consultant to assist with the development of the Hotel.

Development Fee   fee provided to Hotel Manager for supervision of the construction or conversion of significant capital improvement of the Hotel. The Development Fee is either a flat fee, or calculated as a percentage of the total construction costs of the constructing the Hotel. When the Development Fee is calculated as a percentage of total construction costs, the fee is typically a percentage of the hard and soft costs of total construction.

Director of Sales and Marketing   member of the Hotel Executive Personnel who is responsible for sales and marketing of the Hotel.

Dispute Resolution   procedures for settling disputes, such as litigation, arbitration, mediation, or mini-trials. Dispute Resolution procedures not involving litigation are usually less costly and more expeditious, so they are increasingly being used in commercial and labor disputes. Dispute resolution procedures are typically set forth in the Operative Agreement.

Due Diligence   process of confirming material facts. The theory behind Due Diligence is that performing this type of investigation contributes significantly to informed decision making by enhancing the amount and quality of information available to decision makers and by ensuring that this information is systematically used to carefully deliberate on the decision at hand and all its costs, benefits, and risks.

EBITDA   earnings before interest, taxes, depreciation, and amortization. It is also possible to use EBITDA for the calculation of fees. EBITDA means net income (loss) attributable to the hotel, adjusted to exclude the following items, interest expense (net of interest income), income tax expense or benefit, and depreciation and amortization, excluding the effect of extraordinary and/or unusual items, as those items are defined by GAAP in FASB Accounting Standards Codification 225-20. Reserves are also usually excluded.

Effective Date   date on which the Operative Agreement becomes effective.

Electronic System Fee   fee charged by Franchisor to Franchisee for the use of the Franchisor's electronic systems, which are groupings of electronic circuits and components designed to accomplish one or more complex functions, such as tele-communication systems, computer systems, power distribution systems, and electronic music systems.

Emergency Costs   costs and expenses required to (a) correct a condition that if not corrected would endanger imminently the preservation or safety of the Hotel or the safety of tenants, guests or other persons at or using the Hotel, or (b) prevent the Owner or its any of its members from being subjected imminently to criminal or substantial civil penalties or damage.

Employees   *see* Hotel Employees

Employment Laws   any federal, state, local and foreign statutes, laws, ordinances, regulations, rules, permits, judgments, orders, and decrees affecting labor union activities, civil rights, or employment.

Employment Policies   any policies of the employer of the Hotel Personnel in connection with their employment.

Entitlements   all entitlements, zoning, FAR (floor area ratio) and similar approvals required under Applicable Law applicable to the Hotel.

Environmental Laws   all applicable federal, state, municipal and local laws, including all statutes, by-laws and regulations and all orders, directives and decisions rendered by, and policies, instructions, guidelines and similar guidance of, any governmental department or administrative or regulatory agency relating to the protection of the environment, occupational health and safety or the manufacture, processing, distribution, use, treatment, storage, disposal, packaging, transport, handling, containment, clean-up, or other remediation or corrective action of any Hazardous Materials.

Estoppel Certificate   document used to establish facts and financial obligations, such as outstanding amounts due, that can affect any transaction. An Estoppel Certificate is frequently required in Hotel-related transactions to establish or confirm facts directly from the Party issuing the certificate.

Event of Default   a default under the Operative Agreement, as listed in the Operative Agreement, such as (a) the failure of either Party to pay any sum of money to the other Party (or the failure of Owner to provide Working Capital or reimburse any amounts required to be paid or reimbursed by it to Manager or to provide other funds) when due and payable, if such failure is not cured within a short period of time, often five business days after written notice specifying the failure is received by the defaulting Party from the non-defaulting Party, (b) the failure of either Party to perform, keep or fulfill any of the material covenants, undertakings, obligations or conditions set forth in the Operative Agreement, if the failure is not cured within a period of time, such as 30 days after written notice specifying the failure is received by the defaulting from the non-defaulting Party; often subject to the caveat that if (i) the Event of Default is not susceptible of cure within a 30-day period; (ii) the Event of Default cannot be cured solely by the payment of a sum of money; and (iii) the breach or default would not expose the non-defaulting Party to an imminent and material risk of criminal liability or of material damage to its business reputation, the 30-day cure period will be extended if the defaulting Party commences to cure the breach or default within the 30-day period and thereafter proceeds with reasonable diligence to complete the cure and the cure is completed within 120 days after receipt of the notice, (c) the breach by either Party in any material respect of any warranty or representation made in the Operative Agreement, unless the breach will not cause, and is not likely to cause, any material adverse effect on the other Party and does not materially affect the performance by the parties of their respective obligations under the Operative Agreement, (d) the occurrence of an Event of Insolvency with respect to Owner or Manager (in which case the Party suffering the Event of Insolvency will be the defaulting Party), (e) the commission by either Party of fraud or willful misconduct or gross negligence or reckless disregard which does not otherwise constitute a breach of any express covenant or obligation under the Operative Agreement.

Event of Insolvency   with respect to a Party, the occurrence of any one of the following events (a) the Party makes a general assignment for the benefit of creditors, (b) a trustee, custodian or receiver is appointed by any court with respect to the Party or any substantial part of the Party's assets, (c) an action is taken by a Party to seek relief under any bankruptcy or insolvency laws or laws relating to the relief of debtors, or (d) an action is taken against the Party under any bankruptcy or insolvency laws or laws relating to the relief of debtors and such action is not dismissed within a negotiated period of time after commencement of the action.

Executive Personnel   persons selected by Manager from time to time for the senior positions of the Hotel. These positions often include the general manager, director of finance/comptroller, director of sales and marketing, director of revenue management, and director of food and beverage. Typically, the selection of Executive Personnel is subject to the reasonable approval by Owner.

Exclusive Territories   *see* Area of Protection

Expert   independent nationally recognized hospitality industry consulting firm or individual who is qualified to resolve the issue in question, and who is appointed in each instance in accordance with the Operative Agreement.

Extraordinary Event   any of the following events, regardless of where it occurs or its duration, but only to the extent the events impact the operation and performance of the Hotel and only to the extent the events or the consequences of the events were not caused by or within the reasonable control of the Party claiming that the occurrence of the event constitutes an Extraordinary Event: acts of nature without the interference of any human agency (including hurricanes, typhoons, tornadoes, cyclones, other severe storms, winds, lightning, floods, earthquakes, volcanic eruptions, fires, explosions, disease, or epidemics); fires and explosions caused wholly or in part by human agency; acts of war or armed conflict; riots or other civil commotion; terrorism (including hijacking, sabotage, chemical or biological attack, bombing, murder, assault, and kidnapping); strikes or similar labor disturbances; shortage of critical materials or supplies; action or inaction of governmental authorities having jurisdiction over the Hotel (including the imposition of restrictions on room rates or wages or other material aspects of operation, or the revocation or refusal to grant licenses or permits, where the revocation or refusal is not due to the fault of the Party whose performance is to be excused for reasons of the Extraordinary Event); and any other events wholly beyond the reasonable control of a Party, excluding general economic and/or market conditions other than such conditions caused by any of the events described in this definition.

F&B   *see* Food and Beverage

FASB   Financial Accounting Standards Board, a private, not-for-profit organization whose primary purpose is to develop generally accepted accounting principles (GAAP) within the United States in the public's interest. The Securities and Exchange Commission (SEC) designated FASB as the organization responsible for setting accounting standards for public companies in the U.S.

FDD   *see* Franchise Disclosure Document

Federal Trade Commission   an independent agency of the United States government, established in 1914 by the Federal Trade Commission Act. Its principal mission is the promotion of consumer protection and the elimination and prevention of anticompetitive business practices. It regulates franchising, including the franchising of Hotels.

FF&E   *see* furniture, fixtures, and equipment

FF&E Reserve   separate account into which Owner must deposit funds on a monthly basis to have available for the payment of repair and replacement of FF&E.

Fiduciary Duty   duty of a fiduciary to the Party who has entrusted the fiduciary. A fiduciary is a legal or ethical relationship of trust between two or more parties. Typically, a fiduciary prudently takes care of money for another person. For example, a Party who accepts funds from another for safekeeping or investment has traditionally been considered a fiduciary. In a fiduciary relationship, one person, in a position of vulnerability, justifiably vests confidence, good faith, reliance, and trust in another whose aid, advice, or protection is sought in some matter. In such a relation, good conscience requires the fiduciary to act at all times for the sole benefit and interest of the one who trusts the fiduciary.

Final Accounting Statement   accounting statement for the Fiscal Year in which termination or expiration occurs, that is prepared by Manager and delivered to Owner within a specific number of days following the date of termination or expiration of the Operative Agreement.

Fiscal Year a 12-month period starting on a specific date and ending on a specific date during the Term, except that the first Fiscal Year for a Hotel will generally be the period commencing on the Opening Date and ending on December 31 of the year in which the Opening Date occurs. The phrase "full Fiscal Year" means any Fiscal Year containing not fewer than 365 days. A partial Fiscal Year after the end of the last full Fiscal Year and ending with the expiration or earlier termination of the Term generally constitutes a separate Fiscal Year.

Fixed Asset Supplies items included within "Property and Equipment" under the Uniform System of Accounts including, linen, china, glassware, tableware, uniforms, and similar items, whether used in connection with public space or Hotel rooms.

Fixed Charges those charges that are "fixed" and not variable in relation to hotel occupancy, such as real estate and personal property taxes, insurance, deposits into the FF&E Reserve, equipment lease rental and Debt Service.

Fixed Expenses (i) payments pursuant to equipment leases or other forms of financing obtained for the FF&E located in or connected with the Hotel, (ii) rental payments pursuant to any ground lease of the real property on which the Hotel is located, (iii) Impositions, (iv) real estate property and building insurance, and (v) contributions to the Reserve. Fixed Expenses generally do not include debt service payments associated with any Mortgage.

Flag name of the Hotel as associated with the Brand.

Food and Beverage food and beverages offered at the Hotel.

Food and Beverage Operations food and beverage operations at the Hotel.

Foreclosure any exercise of the remedies available to Mortgagee, upon a default under the Mortgage, which results in a transfer of title to, or possession or control of, the Hotel. The term "Foreclosure" includes (i) a transfer by judicial foreclosure; (ii) a transfer by deed in lieu of foreclosure; (iii) the appointment by a court of a receiver to assume possession or control of the Hotel; (iv) a transfer of either Ownership or control of the Owner, direct or indirect in either case, by exercise of a stock pledge or otherwise; (v) a transfer resulting from an order given in a bankruptcy, reorganization, insolvency, or similar proceeding or a transfer approved by a court in such a proceeding; (vi) if title to the Hotel is held by a tenant under a ground lease, an assignment of the tenant's interest in the ground lease; or (vii) a transfer through any similar judicial or non-judicial exercise of the remedies held by the holder of the Mortgage.

Foreclosure Action act of Foreclosure by a Mortgagee.

Foreclosure Date date on which title to, or possession or control of, the Hotel is transferred by means of a Foreclosure after all petitions for rehearing and appeal have been exhausted or the time for filing has expired.

Fractional Franchise Franchise relationship that satisfies the following criteria when the relationship is created: (1) Franchisee, any of the Franchisee's current directors or officers, or any current directors or officers of a parent or Affiliate, has more than two years of experience in the same type of business; and (2) parties have a reasonable basis to anticipate that the sales arising from the relationship will not exceed 20% of the Franchisee's total dollar volume in sales during the first year of operation.

Framework framework of acceptable contract terms which the Owner and Manager establish, prior to commencing or participating in any collective bargaining negotiations, to obtain Owner's direction with respect to the Hotel's negotiating strategy (including the strategic decision as to whether to commit the Hotel to participation in a negotiating group bound to mutual defense and acceptance of the group's collective decision).

Franchise   contractual relationship between the Franchisor and the Franchisee in which the Franchisor offers or is obliged to maintain a continuing interest in the business of the Franchisee in such areas as know-how and training; wherein the Franchisee operates under a common trade name, format or procedure owned by or controlled by the Franchisor, and in which the Franchisee has made or will make a substantial capital investment in its business from its own resources.

Franchise Agreement   legal, binding contract between a Franchisor and Franchisee, included in the FDD, which outlines the responsibilities of both the Franchisor and the Franchisee.

Franchise Disclosure Document   a legal document which is presented to prospective buyers of Franchises in the pre-sale disclosure process in the United States. It was originally known as the Uniform Franchise Offering Circular (UFOC) (or uniform franchise disclosure document), prior to revisions made by the Federal Trade Commission in July 2007. Franchisors were given until July 1, 2008 to comply with the changes. The Federal Trade Commission Rule of 1979 which governs disclosure of essential information in the sale of franchises to the public underlies the state FDD's and prohibits any private right of action for the violation of the mandated disclosure provisions of the FDDs. Therefore, the FDD implies that only the federal government or the state governments have the right to sue and negotiate consent decrees and rescissions with those Franchisors who violate the provisions of the FTC Franchise Rule. However, various state franchise laws that provide for use of an FDD, in lieu of their own disclosure requirements, may create private rights of action, where a franchisor has violated its disclosure obligations in its FDD.

Franchise Fee   initial fee paid to a Franchisor to become a Franchisee, generally outlined in Item 5 of the Franchise Disclosure Document. For some Franchises, this is a flat, one-size-fits-all fee; for others, it varies based on territory size, experience or other factors. Many Franchisors offer franchise fee discounts for veterans, minorities or existing Franchisees.

Franchised   property that is operating under a Franchise.

Franchised Hotel   a Hotel that is operated pursuant to a Franchise Agreement.

Franchisee   an individual who purchases the right to operate a business under the Franchisor's name and system.

Franchisor   parent company that allows individuals to start and run a business using its Trademarks, products, and processes pursuant to a Franchise Agreement.

FTC   *see* Federal Trade Commission

FTC Franchise Rule   Franchise rule which defines acts or practices that are unfair or deceptive in the franchise industry in the United States. The FTC Franchise Rule is published by the Federal Trade Commission. The FTC Franchise Rule seeks to facilitate informed decisions and to prevent deception in the sale of Franchises by requiring Franchisors to provide prospective Franchisees with essential information prior to the sale. It does not, however, regulate the substance of the terms that control the relationship between Franchisors and Franchisees. Also, while the Franchise Rule removed the regulation of the sale of Franchises from the purview of the states under the authority of the FTC to regulate interstate commerce, The FTC Franchise Rule does not require Franchisors to disclose the unit performance statistics of the Franchised system to new buyers of Franchises as would be necessary and material under state and federal Securities and Exchange law.

Furniture, Fixtures, & Equipment furniture, furnishings, wall coverings, fixtures, equipment, and systems located at, or used in connection with the Hotel, together with all replacements and additions, including: all equipment, furnishings, fixtures and systems required for the operation of kitchens, bars, laundry, dry cleaning facilities, and recreational facilities, including Hotel equipment; office equipment; material handling equipment; cleaning and engineering equipment; telephone systems; computerized accounting systems; and vehicles. FF&E will not include any item included in Fixed Asset Supplies or included as part of the Hotel.

GAAP *see* generally accepted accounting principles

General manager general manager of the hotel.

Generally accepted accounting principles generally accepted accounting principles, consistently applied and as modified by the Uniform System and the express terms of the Operative Agreement.

GOP *see* Gross Operating Profit

GOP Test that portion of the Performance Test on which Manager's performance in connection with GOP is measured.

Gross Negligence or **Willful Misconduct** definition varies from state to state. Under New York law, negligence that rises to the level of gross negligence must show reckless indifference to the rights of others. The conduct must show a failure to use even slight care or conduct that is so careless as to show complete disregard for the rights and safety of others. The gross negligence standard focuses on the severity of a Party's deviation from reasonable care. In New York, willful misconduct occurs when a person intentionally acts or fails to act knowing that his or her conduct will probably result in injury or damage. Willful misconduct can also occur when a person acts in so reckless a manner or fails to act in circumstances where an act is clearly required, so as to indicate disregard of his or her action or inaction. A Party claiming willful misconduct must show an "intentional act of unreasonable character performed in disregard of a known or obvious risk so great as to make it highly probable that harm would result." The willful misconduct standard is similar to the gross negligence standard; however, it focuses more on the harm that a Party's action or inaction caused.

Gross Operating Profit all profits from selling any services (e.g., room rates) in a particular period before costs not directly related to producing them, for example, interest payments and tax, are subtracted.

Gross Revenue all revenues and receipts of every kind derived from operating the Hotel and all departments and parts of the Hotel, such as: (i) income (from both cash and credit transactions) from rental of rooms, Food and Beverage Operations, telephone charges, stores, offices, exhibit or sales space of every kind; license, lease and concession fees and rentals, but not including gross receipts of licensees, lessees and concessionaires; (ii) income from vending machines; (iii) net income to the Hotel from parking; (iv) wholesale and retail sales of merchandise; (v) service charges (but not gratuities and service charges paid to Hotel Employees); and (vi) proceeds, if any, from business interruption or other loss of income insurance. Gross Revenues traditionally does not include the following: (a) gratuities and service charges paid to Hotel Employees; (b) federal, state, or municipal excise, sales or use taxes or any other taxes collected directly from patrons or guests or included as part of the sales price of any goods or services; (c) proceeds from the sale of FF&E; (d) interest received or accrued with respect to the funds in the Reserve or the Operating Accounts; (e) any refunds, rebates, discounts and credits of a similar nature, given, paid or returned in the course of obtaining Gross Revenues or components thereof; (f) insurance proceeds (other than proceeds from business interruption or other loss of income insurance); (g) condemnation proceeds (other than for a temporary taking); (h) any proceeds from any sale of the Hotel or from any Mortgage; (i) any and all rental and other revenues derived from the placement of billboards and other signage on the Hotel; (j) proceeds of any judgment or settlement not received as compensation for actual or potential loss of Gross Revenues; (k) any funds contributed or supplied by Owner to the Hotel, including, without limitation, to fund Operating Expenses, Capital Improvements or the Reserve; (l) advances or security deposits from Hotel guests or other Hotel users (other than forfeited deposits); (m) amounts representing the value or cost of room occupancy, meals or other services provided as compensation to Hotel Employees (to the extent permitted by the approved Annual Budget) or as complementary benefits to any other persons; and (n) proceeds of collection of accounts receivable to the extent the receivable was previously included in Gross Revenues.

Gross Room Revenue   with respect to any period, all room revenues actually received by Manager from Guests derived directly from the rental of the Guest Room (including all telephone revenues and including any cancellation fees actually collected by Manager as a result of a reservation designated for a Guest Room having been cancelled in a manner that would entitle Manager to collect the cancellation fees) and properly attributable to the period under consideration determined in accordance with the Uniform System, and determined on the accrual method of accounting, less and except: (i) any applicable excise, value added, sales, use, occupancy, bed, resort, tourism and/or other similar government taxes, duties, levies or charges assessed in conjunction with the renting of the Guest Room; and (ii) any Incidental Charges.

Gross Up   all payments must be made in the full amount, free of any deductions or withholdings, and without exercising any right of set-off. The gross-up provision in the Operative Agreement will usually indicate that if there is a mandatory withholding or deduction by operation of law (usually with respect to a tax), then the paying Party must increase or "gross up" the payment so that the receiving Party receives the same net amount.

Guarantor   a person who promises to perform certain obligations of another. This would include payment and performance obligations under a Management Agreement, Franchise Agreement or if the Borrower defaults under the Loan Documents.

Guaranty   legal document that contains the warrant, pledge, or formal assurance given by a Guarantor to secure the obligations of another.

Guest   a member of the general public who is a guest of the Hotel, either by occupying a Guest Room, or transacting other lawful business at the Hotel.

Guest-Facing Amenities   those amenities offered at the Hotel to enhance the Guest's stay and experience at the Hotel are personal and unique to the Guest. What is considered an "amenity" changes over time, from market to market, and from asset class to asset class of the hotel industry. Amenities are often the differentiator of one comparable Hotel to another when making a choice between similarly situated hotels. The focus on Guest-Facing Amenities is particularly emphasized in the luxury segment and the Boutique Hotel market.

Guest Revenues   *See* Gross Revenue

Guest Room   each rentable unit in the Hotel consisting of a room or suite of rooms generally used for overnight guest accommodations, entrance to which is controlled by one key. Adjacent rooms with connecting doors that can be locked and rented as separate are deemed to be separate guest rooms.

Hazardous Materials   any substance or material identified by any Legal Regulations as being hazardous to the health and safety of guests, Hotel Employees, or any Person in connection with the Site and requiring the monitoring, clean-up, or removal of such substance. Hazardous Materials shall include, without limitation, asbestos, lead-based paint, and PCBs.

HMA   *see* Hotel Management Agreement

Hotel   each element and component of certain hotel that is typically described in detail in the recitals of the Operative Document, and shall include the Site together with the Building and all other improvements constructed or to be constructed on the Site pursuant to the Operative Agreement, all FF&E and Fixed Asset Supplies installed or located on the Site or in the Building, and all easements or other appurtenant rights thereto.

Hotel Employees   all personnel employed at the Hotel.

Hotel Guest Data   Guest information of the Hotel, including, without limitation, the names, addresses, email addresses, and other contact information for all guests of the Hotel, and all related guest stay history, but excluding (i) any qualitative analysis of any such guest history data and any guest profiles created by Manager using such data, and (ii) any data independently generated through the System that is not directly related to guest information of the Hotel.

Hotel Guest Records   *see* Hotel Guest Data

Hotel Management Agreement   management agreement, between Owner and Manager pursuant to which Manager will manage the Hotel. The term "Hotel Management Agreement" includes (i) any amendments, modifications, supplements, replacements or extensions of the original Management Agreement; and (ii) any and all other agreements entered into with respect to the management and operation of the Hotel.

Hotel Manager   manager of the Hotel and includes any permitted successor or assign, as applicable.

Hotel Operating Expenses   *see* Operating Expenses

Hotel Owner   *see* Owner

Hurdle   either a fixed amount or a methodology for determining an amount that is mutually agreed upon by Owner and Manager that must be met before Manager is entitled to an Incentive Management Fee.

IMF   *see* Incentive Fee

Impositions   all real estate and personal property taxes, levies, assessments and similar charges on or relating to the Hotel or attributable to the Hotel or any of its component parts.

Imputed Liability   attaching responsibility to a person (including financial liability) for acts or injuries to another, because of a particular relationship, such as an employer and employee relationship.

Incentive Fee   fee that is paid to Hotel Manager for the incremental profitability of the Hotel due to Hotel Manager's operational expertise. Typically, an Incentive Fee is earned by the Hotel Manager after the Owner receives a return of a certain portion of its investment in the Hotel. This is a matter of negotiation, but could be limited to invested equity only or debt and equity as well as additional capital invested during the Term for Capital Improvements. The Incentive Fee is due monthly, quarterly or annually, subject to a true-up at the end of the Fiscal Year. The structure of the incentive fee can be complex and is often heavily negotiated in order to meet the specific requirements of both the Owner and Hotel Manager. Also known as Incentive Management Fee.

Incentive Management Fee   *see* Incentive Fee

Incidental Charges   revenue generated from any source other than Gross Room Revenue and any other fees or charges to a Guest occupying the Guest Room, including, without limitation, minibar purchases, pay-per-view television services, food and beverage purchases, internet access charges, business center charges, show tickets or other activities, amenities or other sales or service products provided by Manager or its designees, charges for use of meeting space (if any), dry cleaning services, valet parking services, and other fees and charges related to other services offered by the Hotel.

In-House Financing   financing offered by the Franchisor to Franchisees to help with expenses, which can include the initial Franchise Fee, startup costs, equipment and inventory as well as day-to-day expenses such as payroll.

Initial Public Offering   corporation's first offer to sell stock to the public.

Initial Supplies   supplies that must be on hand at the Hotel prior to the Opening Date.

Initial Term   initial term of the Operative Agreement prior to the exercise of any renewal options.

Intellectual Property   those trade names, trademarks, service marks, trade dress, copyrights, logos, insignia, emblems, symbols, slogans, distinguishing characteristics, domain names, or other indicia of origin, and other identifying characteristics of the Brand associated or used in connection with the System, and specifically excluding any of Licensor's or its Affiliates' other trademarks, trade names, brands, or product lines owned or licensed or being registered by Licensor or its Affiliates and identifying brands other than the Brand.

Interim License Agreement   agreement whereby the existing holder of the Liquor License agrees to enter into an interim arrangement with Buyer or its designee for a specified period of time after Closing.

Interior Design Consultation/Review Services   those design and consultation services in connection with the interior of the Hotel and the FF&E, to ensure that the design of the Project meet the Standards of the Brand.

International Services Agreement   agreement whereby Licensor provides services to Owner to be performed in support of the Hotel outside the country in which the Hotel is located.

Inventoried Baggage   inventory of all baggage and similar items left in the care of Manager at the Hotel and all lost and found items belonging to Guests.

Inventories   as defined in the Uniform System, all items such as provisions in store-rooms, refrigerators, pantries and kitchens; beverages in wine cellars and bars; other merchandise intended for sale; fuel; mechanical supplies; stationery; and other expensed supplies and similar items.

Key Personnel certain Hotel Employees who hold senior executive positions at the Hotel such as General Manager, Chief Financial Officer and Director of Sales and Marketing.

Legal Requirements any federal, state or local law, code, rule, ordinance, regulation or order of any governmental authority or agency having jurisdiction over the business or operation of the Hotel or the matters that are the subject of the Operative Agreement, such as (i) any building, zoning or use laws, ordinances, regulations or orders, and (ii) all applicable Environmental Laws.

Letter of Intent non-binding agreement that sets forth the key terms of the Operative Agreement as agreed upon by the Parties prior to preparation of the Operative Agreement.

License Agreement agreement pursuant to which Licensor grants Licensee certain rights to use the Brand.

License Fee fee paid for the use of the Brand's intellectual property and name.

Licenses and Permits all necessary operating licenses and permits, including liquor, bar, restaurant, sign, and Hotel licenses, as may be required for the operation of the Hotel.

Licensing and Royalty Agreement agreement whereby Licensor grants to Owner a non-exclusive and non-transferable license to use the Trademarks for hotel services in a particular location and only in connection with the operation of the Hotel under the Operative Agreement.

Licensor Party which has the right to License the Brand to the Licensee.

Lifestyle Hotel a Hotel that combines living elements and activities into functional design that gives Guests the opportunity to explore the experience they desire.

Liquidated Damages an agreed upon amount to be paid by one Party to another as agreed upon damages and not as a penalty in the event of a default under the Operative Agreement.

Liquor License the alcoholic beverage and beer licenses issued by the governing authority in connection with the Hotel.

Litigation (i) any cause of action (including, without limitation, bankruptcy or other debtor/creditor proceedings) commenced in a federal, state, or local court, and (ii) any claim brought before a governmental administrative agency or body.

Loan act of giving money, property, or other material goods to another Party in exchange for future repayment of the principal amount along with interest or other finance charges. A Loan may be for a specific, one-time amount or can be available as open-ended credit up to a specified ceiling amount.

Loan Acceleration mechanism that allows a Mortgagee to require a Borrower to repay all or part of an outstanding Loan if certain requirements are not met. A loan acceleration clause outlines the reasons that the Mortgagee can demand Loan repayment.

Loan to Cost ratio used in commercial real estate construction to compare the amount of the Loan used to finance the Hotel to the cost to build the Hotel.

Loan Documents documents the Mortgagee requires a Borrower to sign before the Mortgagee will advance Loan monies to the Borrower.

Loan to Value financial term used by Mortgagees to express the ratio of the Loan amount to the value of the asset being financed.

Lock Out Period a predetermined amount of time following the Effective Date of the Hotel Management Agreement where the Owner is prohibited from exercising its right of Termination Upon Sale.

LOI   *see* Letter of Intent

Losses   any and all costs, fees, expenses, damages, deficiencies, interest, and penalties (including, without limitation, reasonable attorneys' fees and disbursements) suffered or incurred by a Party.

LTC   *see* Loan to Cost

LTV   *see* Loan to Value

Major Purchases   those purchases that arise in connection with a substantial repair or renovation of the Hotel.

Manage   act of managing the Hotel.

Managed Property   a Hotel operating under a Hotel Management Agreement and/or a License Agreement rather than a Franchise Agreement.

Management Agreement   *see* Hotel Management Agreement

Management Fees   base management fee together with the incentive management fee.

Manager   manager of the Hotel and includes any permitted successor or assign, as applicable.

Market Risks   possibility for a Party to experience losses due to factors that affect the overall performance of the hospitality and financial markets. "Market risk," also called "systematic risk," cannot be eliminated through diversification, although it can be hedged against. The risk that a major natural disaster will cause a decline in the market as a whole is an example of market risk. Other sources of market risk include recessions, political turmoil, changes in interest rates and terrorist attacks.

Marketing Budget   a budget for advertising and marketing the Hotel, including proposed discount and complementary policies, detailing, on a line–item basis, the costs associated with the Marketing Plan.

Marketing Fee   a fee to compensate the Hotel Manager for all advertising, publicity, and sales and marketing services. The Marketing Fee is fairly common, especially in the context of a Branded Hotel, either as a standalone fee, or incorporated in the Centralized Services Fee.

Marketing Fund   a specific fund set aside in connection with the Marketing Budget.

Marketing Plan   a reasonably detailed program for advertising and marketing the Hotel in accordance with the Marketing Budget.

Marks   any specific word associated with the Brand and all other words, trademarks, service marks, trade names, symbols, emblems, logos, insignias, indicia of origin, slogans, and designs (including restaurant names, lounge names, or other outlet names) used or registered by Licensor/Franchisor and which are used to identify or are otherwise used in connection with Branded Hotels, private clubs, timeshare resorts, residential properties or other facilities operated under the Brand name (whether registered or unregistered and whether used alone or in connection with any other words, trademarks, service marks, trade names, symbols, emblems, logos, insignias, indicia of origin, slogans, and designs), all as may be amended, modified, deleted or changed by Licensor.

Master Franchisee   a franchisee who serves as a sub-franchisor for a certain territory. Master Franchisees can issue FDDs, sign up new Franchisees, provide logistical support, and receive a portion of the territory's royalties.

Memorandum of Management Agreement   a memorandum recorded in the land records where the Hotel is located which sets forth key terms of the Hotel Management Agreement.

Memorandum of Understanding   *see* Letter of Intent

Mezzanine Mortgage   a Mortgage reflecting a hybrid of debt and equity financing.

Mezzanine Mortgagee   a Mortgagee pursuant to a Mezzanine Mortgage.

Minimum Return Test   an agreed upon minimum return which Manager must deliver to Owner for a certain number of years. In the event that Manager does not meet the Minimum Return Test, Owner has the right to terminate Manager.

Monthly Report   detailed financial information for the immediately preceding month in form and substance consistent with Manager's commercially reasonable reporting practices, together with any other information required under the Operative Agreement or reasonably requested by Owner.

Mortgage   any mortgage that encumbers the Hotel. The term "Mortgage" includes (i) any amendments, modifications, supplements, or extensions of the original "Mortgage" and (ii) any existing or future financing by Mortgagee that is wholly or partially secured by the Hotel, including a "blanket mortgage," encumbering properties other than the Hotel.

Mortgagee   any of the following: (i) the entity identified as the "Mortgagee" in a Mortgage (ii) any successors or assigns of that entity, (iii) any nominee or designee of that entity (or any other entity described in this definition), (iv) any initial or subsequent assignee of all or any portion of the interest of that entity in the Mortgage, or (v) any entity that is a participant in the financing secured by the Mortgage, or otherwise acquires an equitable interest in the Mortgage.

Mortgagor   an entity that borrows money by mortgaging its property to the Mortgagee as security for a loan.

Mystery Shopper   a representative of the Brand who is sent to the Hotel to test whether the Hotel is meeting Brand Standards.

National Labor Relations Act   an act of Congress (1935) that forbade any interference by employers with the formation and operation of labor unions.

National Labor Relations Board   a board consisting of five members, originally established under the National Labor Relations Act to guarantee workers' rights to organize and to prevent unfair labor practices.

National Sales Program   national sales program for the System which is operated in a fair and equitable manner to benefit all hotels in the System, and whereby no hotels (including hotels owned in whole or part by Manager or its Affiliates) will receive a disproportionate benefit from or be allocated a disproportionate cost of Brand Marketing Services and National Sales Programs.

Net Operating Income   for any period, the amount, if any, by which Hotel Gross Operating Profit exceeds Fixed Expenses.

NLRB   *see* National Labor Relations Board

NOI   *see* Net Operating Income

NOI Deductions   deductions from Net Operating Income.

Offshore Management Agreement   a certain agreement whereby Owner engages Offshore Manager for the purpose of formulating general, overall, and strategic management policies for the Hotel, which policies and day-to-day onsite management will be implemented by the Onshore Manager, an Affiliate of Offshore Manager, pursuant to the Onshore Management Agreement.

Offshore Manager   Manager pursuant to the Offshore Management Agreement.

Onshore Management Agreement   that certain agreement whereby Owner engages Onshore Manager to be responsible for the day-to-day management of and operation of the Hotel.

Onshore Manager   manager pursuant to the Onshore Management Agreement.

Opening Date   date on which the Hotel opens to the general public. Typically, the Parties set forth parameters on which the Opening Date will occur such as the date on which both (i) the Hotel has at least a minimum percentage or number of Guest Rooms open and available to the general public for occupancy by paying guests and (ii) the temporary or final certificate of occupancy of the Hotel has been received, or, if requested by Manager, a later date that is, in any event, within 21 days of when clause (i) and (ii) have occurred.

Operating Account   primary account or accounts used in the operation of the Hotel. The Operating Accounts are usually designated by Owner and are in the name of both Owner and Manager. The Operating Accounts are typically under the control of Manager, as agent and fiduciary of Owner, but in all other respects remain the property of Owner. Withdrawals from the Operating Accounts are typically made solely by representatives of Manager whose signatures have been authorized and are bonded or insured in accordance with the Operative Agreement.

Operating Budget   *see* Annual Budget

Operating Cash Flow   a measure of the amount of cash generated by the Hotel's normal business operations. Operating cash flow is important because it indicates whether the Hotel is able to generate sufficient positive cash flow to maintain and grow its operations, or whether it may require external financing, such as through additional cash from Owner to fund operating shortfalls.

Operating Expenses   all expenses incurred by Manager on behalf of Owner in operating the Hotel in accordance with and subject to the approved Annual Budget, the terms of the Operative Agreement or otherwise approved in advance by Owner. A typical schedule of Operating Expenses is likely to include: (1) the cost of sales, including compensation, fringe benefits, payroll taxes, all out-of-pocket and travel expenses to the Hotel for Hotel Employees providing services to the Hotel, and any employment costs of Corporate Personnel allocable to the Hotel; (2) departmental expenses incurred at departments within the Hotel; administrative and general expenses; the cost of marketing incurred by the Hotel; advertising and business promotion incurred by the Hotel; heat, light and power; computer line charges; and routine repairs, maintenance and minor alterations treated as Operating Expenses; (3) the cost of Inventories and Fixed Asset Supplies consumed in the operation of the Hotel; (4) a reasonable reserve for uncollectible accounts receivable as determined by Manager; (5) all costs and fees of independent professionals or other third parties who are retained by Manager to perform services required or permitted hereunder; (6) all costs and fees of technical consultants and operational experts who are retained or employed by Manager and/or Affiliates of the Manager for specialized services (including, without limitation, quality assurance inspectors) and the cost of attendance by Hotel Employees at training and manpower development programs sponsored by Manager; (7) the Base Management Fee, the Brand Marketing and National Sales Contribution, the Reservation Fee, and all Reimbursable Expenses, if any; (8) insurance costs and expenses, other than costs and expenses of property and building insurance (which is generally expressed as a Fixed Expense); (9) taxes, if any, payable by or assessed against Manager related to the Operative Agreement or to Manager's operation of the Hotel (exclusive of Manager's income taxes); and (10) such other costs and expenses incurred by Manager as are specifically provided for elsewhere in the

Operative Agreement, the approved Annual Budget or are otherwise approved in advance by Owner.

Operating Loss   a negative Gross Operating Profit.

Operating Standards   level of service and quality generally considered to be in accordance with the market sector of the Hotel and no less than the level of service and quality prevailing from time to time at the Brand Hotels, consistent with the Brand Standards, and in accordance with the Operative Agreement.

Operating Supplies and Equipment   small equipment which is manufactured offsite and most of if it does not require any installation, such as glassware, chinaware, silverware, table cloths, banquet furniture, lecterns, portable dance floor, staging, audio visual, bins, TVs, minibars, room safes, bed sheets, towels, kitchen utensils, trolleys, vacuum cleaners, shelving, refuse bins, lockers, uniforms, desks, and chairs. Operating Supplies and Equipment does not include consumable supplies such as food, drink, or paper products.

Operating Year   a 12-month period commencing on a certain date and terminating 12 months thereafter, as agreed on by Owner and Manager, in which the Owner and Manager operate the Hotel and measure the performance of the Hotel.

Operative Agreement   for purposes of this glossary, the certain agreement in which a certain definition would be used.

Operator   manager of the Hotel and includes any permitted successor or assign, as applicable. Also known as Manager, Hotel Manager, or Hotel Operator.

OS&E   *see* Operating Supplies and Equipment

Outside Commencement Date   *see* Outside Opening Date

Outside Opening Date   a certain date agreed on by the Parties by which the Opening Date of the Hotel must have occurred. If the Opening Date has not occurred on or before the Outside Opening Date, Consultant/Manager/Licensor may have the right to terminate the Operative Agreement.

Owner   *see* Hotel Owner

Owner's Investment   generally, the sum of (i) total hard and soft costs required to develop and construct the hotel, such as (a) design, architecture, engineering, legal and similar "soft costs," (b) construction and similar "hard" costs, (c) the cost of the initial FF&E, fixed asset supplies and inventories, (d) tenant improvement costs paid by owner from time to time, if any, (e) construction period interest on any construction loan for the hotel, (f) pre-opening costs and expenses, (g) initial working capital for the operation of the hotel, and (h) the fair market value of the land underlying the site, *plus* (ii) the cost of any Capital Improvements that exceed any funds available for Capital Improvements in the Reserve, funded by Owner from other than the Reserve, regardless of how Owner funds the Capital Improvements.

Owner's Priority   an agreed on return on investment for Owner, usually as a prerequisite to Manager's Incentive Management Fee.

Owner's Priority Return   *see* Owner's Priority

Owner's Remittance Amount   payment to Owner of the amount by which the total funds then in the Operating Account exceeds the Working Capital to be maintained in the Operating Account.

Owner's Total Investment   typically, all sums expended to acquire, develop, construct and finance the Hotel, including hard and soft costs and loan proceeds. Following the opening of the Hotel, Owner's Total Investment should include all sums expended for capital improvements beyond the sums funded into the Reserve and, depending on the Owner's bargaining power, the initial and any subsequently funded Working Capital. The definition of Owner's Total Investment often includes the acquisition and development loan proceeds amount and actual construction period Debt Service (or Debt Service based on a predetermined "base constant" rate) during the construction period. Owner's Total Investment and Owner's Investment are often the same and only one of the defined terms appears in the Operative Agreement.

Ownership   ownership of the Hotel.

Party   a Party to the Operative Agreement.

Perfected Security Interest   perfection as it relates to the additional steps required to be taken in relation to a security interest in order to make it effective against third parties and/or to retain its effectiveness in the event of default by the grantor of the security interest. Generally speaking, once a security interest is effectively created, it gives certain rights to the holder of the security and imposes duties on the Party who grants that security.

Performance Failure   failure of Operator to meet the Performance Test.

Performance Termination Notice   termination notice provided by Owner to Operator in the event of a Performance Failure.

Performance Test   test agreed on by the Parties to measure the Operator's performance under the Hotel Management Agreement. For example, Owner may terminate the Hotel Management Agreement, if, for any two consecutive Performance Test Periods, the operations of the Hotel fail to achieve (i) the agreed upon percentage of the amount of the Hotel Gross Operating Profit set forth in the approved Annual Budget for each of such Performance Test Periods (the "Operating Profit Test"), and (ii) the agreed upon percentage of the mutually agreed upon RevPAR Index of the Competitive Set (the "RevPAR Test") for each of such Performance Test Periods.

Performance Test Commencement Date   date on which the Performance Test commences as agreed upon by the Parties.

Performance Test Financial Statement   Annual Operating Statement for the Performance Test Period.

Performance Test Period   agreed on test period during the Term commencing on the Performance Test Commencement Date.

Person   an individual (and the heirs, executors, administrators, or other legal representatives of an individual), a partnership, a corporation, limited liability company, a government or any department or agency thereof, a trustee, a trust and any unincorporated organization.

PIP   *see* Property Improvement Plan

POTSA   *see* Pre-Opening and Technical Services Agreement

Pre-Opening Budget   budget pursuant to which pre-opening services are provided in accordance with the Pre-Opening Services Agreement, or the Pre-Opening and Technical Services Agreement, as applicable.

Pre-Opening Expenses   costs and expenses in connection with all initial Inventories, Pre-Opening Human Resources, advertising, marketing and public relations expenses as set forth in the Inventories Budget, Pre-Opening Marketing Budget, Pre-Opening Human Resources costs, payments made by Manager or its Affiliates, employees, or consultants to third parties for goods and services in connection with the Hotel, and out-of-pocket-expenses incurred by Manager in connection with its performance of the Pre-Opening Services, all to the extent provided for in the Pre-Opening Budget or as otherwise authorized under the Operative Agreement.

Pre-Opening Human Resources   Hotel Employees required during the Pre-Opening Period.

Pre-Opening Marketing   marketing of the Hotel during the Pre-Opening Period.

Pre-Opening Services Agreement   agreement by and between Owner and Consultant whereby Consultant provides Pre-Opening Services in connection with the opening of the Hotel. This does not include the provision of Technical Services.

Pre-Opening Services Fee   a fee which compensates the Manager for the services of Manager prior to the Opening of the Hotel, such as the preparation and implementation of a pre-opening plan and budget, which will include a preliminary staffing plan, budget, and employee training programs. Typically, the Pre-Opening Services Fee is a flat fee paid monthly to Manager during the Pre-Opening Period.

Pre-Opening Technical Services Agreement   agreement by and between Owner and Consultant whereby Consultant provides pre-opening and technical services in connection with the opening of the Hotel.

Project   project that is the subject of the Operative Agreement.

Property   property being sold and conveyed by Seller and purchased by Purchaser pursuant to the Purchase and Sale Agreement, such as the land, improvements on the land, intangible property, tangible property, personal property, Inventories, FF&E, OS&E, Licenses and Permits, warranties, operating agreements, equipment leases, bookings and reservations, guest data, inventories, and retail merchandise.

Property Improvement Plan   a document detailing the property upgrades and replacements that will be required if a Hotel is to be accepted as one of a specific Brand's Franchised properties.

Proprietary Marks   *see* Trademarks

Proprietary Software   all computer software and accompanying documentation (including all future upgrades, enhancements, additions, substitutions, and modifications thereof) that are proprietary to Manager, its Affiliates, or the System and which is used exclusively by Manager or its Affiliates in connection with operating the System, including any proprietary property management system, reservation system and other electronic systems used by Manager in connection with operating or otherwise providing services to the System and the hotels within the System from time to time.

PSA   *see* Purchase and Sale Agreement

Purchase and Sale Agreement   agreement pursuant to which the Purchaser acquires the Property and the Seller sells the Property to Purchaser.

Purchaser   *see* Buyer

Purchasing Services   services of Manager's corporate purchasing personnel and facilities or an independent third-party purchasing agent for the purchase of Fixed Asset Supplies, FF&E, and other goods and services for the operation of the Hotel.

Quarterly Report   detailed financial information for the immediately preceding calendar quarter in form and substance consistent with Manager's commercially reasonable reporting practices, together with any other information required under the Operative Agreement or reasonably requested by Owner.

Radius Restriction   *see* Area of Protection

Real Estate Investment Trust   a corporate entity that owns income-producing real estate. REITs own many types of commercial real estate, ranging from office and apartment buildings to warehouses, hospitals, shopping centers, hotels, and even timberlands. Some REITs also engage in financing real estate. The REIT structure was designed to provide a real estate investment structure similar to the structure mutual funds provide for investment in stocks. REITs can be publicly or privately held. Public REITs may be listed on public stock exchanges.

Rebate   a rebate, payment, training allowance, credit or other enrichment received by Manager or its Affiliate with respect to or which is measured by the purchase, sale, lease or other procurement or provision of goods, services, systems and/or programs at one or more hotels managed by Manager (including the Hotel) or specifically for the Hotel.

Receiver   a court appointed Party authorized to operate a hotel, company or asset in a manner intended to help preserve the value of the asset.

Recognition Agreement   an agreement among multiple Parties to recognize the rights of other Parties in certain circumstances, such as recognizing the rights of the Manager under a Management Agreement after an Owner default under the Loan Documents.

Registration States   states that require Franchisors to register their FDDs with a state agency before they are legally allowed to sell Franchises within that state.

Reimbursable Expenses   all travel, lodging, meals, entertainment, telephone, tele-copy, postage, courier, delivery, and other expenses incurred by Manager which are related to Manager's performance of the Hotel Management Agreement and are incurred by Manager in accordance with the approved Annual Budget, are other-wise expressly permitted by the Management Agreement or are otherwise approved in advance by Owner, without mark-up by or profit to Manager or its Affiliate. The Management Agreement may go on to specify expenses that will not constitute Reimbursable Expenses and accordingly would not be reimbursed by Owner and would be borne exclusively by Manager or its Affiliate from its own funds. A representative schedule of those expenses might include: (i) all expenses, salaries, wages, or other compensation of Corporate Personnel and any expenses of corporate, regional, principal or branch offices of Manager and its Affiliates or of any other hotel owned or managed by Manager or its Affiliates; (ii) any travel expenses of Corporate Personnel that exceed the amounts permitted to be recovered by Hotel Manager pursuant to the approved Annual Budget; and (iii) any interest or penalty payment with respect to an imposition or lien on the Hotel or Owner by reason of the failure of Manager to make a payment required to be made by Hotel Manager under the Hotel Management Agreement.

REIT   *see* Real Estate Investment Trust

Renewal Option   option for the Parties to extend the Term of the Operative Agreement beyond the Initial Term.

Renewal Period   *see* Renewal Term

Renewal Term   extension of the Initial Term.

Rental Management Agreement   that certain document pursuant to which the Operator manages the rental of Hotel Units.

Repairs and Equipment Estimates   amount the Manager estimates Owner will have to expend to maintain and operate the Equipment.

Reservation Fee   a fee generated from the processing of reservation calls from a call center and internet reservations and reimbursement of certain operational expenses from reservations. This fee can be incorporated into the Centralized Services Fee at the election of the Manager.

Reservation Services   services provided to Owner by Manager in connection with the reservations at the Hotel, and is intended to include all manner of electronic and voice reservation services provided by Manager and/or its Affiliates through telephone reservations arranged through call centers, regardless of where the call center is located, reservations through the internet, or through any number of Global Distribution Systems such as Amadeus/System One, Apollo/Galileo, Sabre (Abacus), and Worldspan, which may involve additional charges for these systems.

Reservation System   system pursuant to which Manager provides Reservation Services to Owner.

Reserve   *see* FF&E Reserve

Residences   certain Branded Residences associated with the Hotel.

Residences Management Agreement   agreement pursuant to which Manager manages the Residences.

Residences Marketing and Licensing Agreement   agreement pursuant to which Manager or its Affiliate manages the Branded Residences and licenses the use of the Brand to the Owner or an association of unit owners.

RevPAR   a performance metric in the hotel industry which stands for "revenue per available room."

RevPAR Index   measure of how an individual Hotel is performing in RevPAR against its Competitive Set. It is calculated by dividing hotel RevPAR by the RevPAR of the Competitive Set. A RevPAR Index below 100 means the market is outperforming the Hotel; whereas a RevPar Index over 100 means that the Hotel is outperforming the Competitive Set.

RevPAR Index of Competitive Set   Revenue Per Available Room of the Competitive Set to establish an index to measure the performance of the subject Hotel. These data are included in a report that is currently produced by Smith Travel Research. Most Management Agreements go on to say that if Smith Travel Research is no longer in existence at any time during the Term, then the report produced by the successor of Smith Travel Research or such other industry resource that is equally as reputable as Smith Travel Research will be substituted, in order to obtain substantially the same result as would be obtained if Smith Travel Research had not ceased to be in existence.

RevPAR Test   that portion of the Performance Test upon which Manager's performance in connection with RevPAR is measured.

Right of First Negotiation   *see* Right of First Offer

Right of First Offer   a contractual obligation by the Owner to a Manager to negotiate the sale of the Hotel with the Manager before offering the Hotel for sale to third parties. If the Manager is not interested in purchasing the Hotel or cannot reach an agreement with the Owner, the Owner has no further obligation to the Manager and may sell the Hotel within the pricing or other parameters specified in the Management Agreement.

Right of First Opportunity to Purchase   *see* Right of First Offer

Right of First Refusal   a contractual right that gives Manager the option to purchase the Hotel from the Owner, according to specified terms, before the Owner is entitled to sell the Hotel to a third party. A Right of First Refusal differs from a Right of First Offer (ROFO) in that the ROFO merely obligates the Owner to undertake exclusive good faith negotiations with the Manager before negotiating with other parties. A ROFR is a first option for the Manager to enter a transaction on exact or approximate transaction terms before the Owner takes those terms to the market. Also known as a Right of First Negotiation.

RMA   *see* Rental Management Agreement

ROFO   *see* Right of First Offer

ROFR   *see* Right of First Refusal

Routine Maintenance, Repairs, and Minor Alterations   routine or non-major repairs and maintenance to the Building which are normally not capitalized (and instead are expensed) under Generally Accepted Accounting Principles, such as exterior and interior repainting and resurfacing building walls, floors, roofs and parking areas, and replacements and renewals related to the FF&E at the Hotel.

Royalty Fee   fee paid by Franchisee to Franchisor on a regular basis, usually monthly. Usually, it is a percentage of Gross Rooms Revenue; sometimes it is a flat fee.

Sale of the Hotel   any sale, assignment, transfer or other disposition, for value or otherwise, of the fee simple title or leasehold estate to the Site and/or the Hotel, other than a Foreclosure or the conveyance of the Site pursuant to a ground lease where Owner or its Affiliate is tenant thereunder. For purposes of many Operative Agreements, a Sale of the Hotel also includes a lease (or sublease) of all or substantially all of the Hotel and any sale, assignment, transfer or other disposition, for value or otherwise, in a single transaction or a series of transactions, of the Controlling Interest in Owner.

Sale Termination Fee   any agreed on Termination Fee paid by Owner to Operator in the event of Termination Upon the Sale.

SBP   *see* Statement of Basis and Purpose

Seller   Party selling the Hotel or an interest in the Owner of the Hotel.

Settlement Statement   a document which sets forth a line-by-line breakdown of costs involved in a real estate sale. This document lists all the costs involved in the sale. Seller and Buyer provide data for the preparation of the Settlement Statement, but it is typically prepared by the title insurance company or escrow agent responsible for conducting the closing of the transaction.

Site   the land on which the Hotel is located, the legal description of which is typically set forth on Exhibit A of the Operative Agreement.

Smith Travel Research   private company that gathers information and acts as a clearing house and amalgamator and organizer of operating data (such as RevPAR, ADR, and occupancy). Smith Travel Research then aggregates this information with data from other hotels in the same market and allows participating hotels to compare themselves to their competition.

SNDA   *see* Subordination Non-disturbance and Attornment Agreement

Sophisticated Franchisee Exemption   an exemption applicable to entities (including any parent or Affiliates) that have been in business for at least five years and have a net worth of at least $5 million. A large Franchisee need not have five years of business experience in franchising or in the industry that the Franchisee will enter as a result of the Franchise; five years of business experience in any business will suffice.

SPE   *see* Special Purpose Entity

Special Purpose Entity   also referred to as a "bankruptcy-remote entity" whose operations are limited to the acquisition and financing of specific assets. The SPE is usually a subsidiary company with an asset/liability structure and legal status that makes its obligations secure even if the parent company goes bankrupt.

Specially Designated National an individual or organization on the Specially Designated Nationals (SDN) List of the Office of Foreign Assets Control, which lists individuals and organizations with whom United States citizens and permanent residents are prohibited from doing business.

Spring Back Provision a provision in the Management Agreement which provides that, in the event Manager was wrongfully terminated by Owner, or if Manager was precluded from operating the Hotel in a manner not contemplated by the Management Agreement, then Manager's right to manage the Hotel under the Management Agreement will "spring back" when a subsequent Owner acquires the Hotel through Foreclosure.

Springing Franchise conversion of a Hotel from a Managed Property to a Franchised Property.

Stabilization Period that negotiated period of time after the Opening Date for the stabilization of the Hotel. Owner is precluded from terminating the Manager due to Performance Failure during this period of time.

Standards *see* Brand Standards

Startup Administration Fee administrative fee paid by Franchisee to Franchisor for the administrative services required at the startup of the Franchise in connection with a Franchised Hotel.

Startup Cost/Initial Investment total amount required to open the Franchise, outlined in the FDD. This includes the Franchise Fee, along with other startup expenses such as the cost of real estate acquisition, equipment, supplies, business licenses, and Working Capital.

Statement of Basis and Purpose that document which accompanied the FTC Franchise Rule, "the basis for the large investment exemption is not that 'sophisticated' investors do not need pre-sale disclosure, but that they will demand and obtain material information with which to make an informed investment decision regardless of the application of the FTC Franchise Rule."

STR Report a report issues by Smith Travel Research (STR).

Subordination Non-disturbance and Attornment Agreement agreement among Owner, Manager, and Mortgagee, whereby, (1) Manager agrees to subordinate its Hotel Management Agreement and other interests in and to the Hotel to Mortgagee's lien, (2) Mortgagee typically agrees not to disturb Manager's enjoyment and control of the Hotel, and not to attempt to terminate the Hotel Management Agreement executed by the Owner/borrower or to remove Manager, and (3) Manager agrees to recognize Mortgagee, or its successor in interest, as the new owner after Mortgagee forecloses or acquires the Hotel by deed in lieu of Foreclosure.

Subsequent Owner any individual or entity that acquires title to, or assumes or obtains possession or control of, a Hotel at or through a Foreclosure (together with any successors or assigns), including: (i) Mortgagee, (ii) any purchaser of the Hotel from Mortgagee, or any lessee of the Hotel from Mortgagee, (iii) any purchaser of the Hotel at Foreclosure, or (iv) any receiver appointed by a court to assume possession or control of the Hotel.

Survival Period that period of time during which certain provisions survive after Closing for example (a) representations, warranties and indemnifications contained in the Operative Agreement may survive for a specific number of months after Closing, and (b) the covenants contained in the Operative Agreement may survive for so long as the covenant remains in effect.

System   a proprietary system for providing a hotel facility with the standard of service, courtesy and cleanliness identified with the Brand and other programs and procedures described in the Brand manuals or otherwise contemplated by the Operative Agreement, in each case, to the extent the same distinguishes and is unique to Brand Hotels, as compared to similar services provided by any other Person.

System Services   those services as part of the System that are provided by Hotel Manager or a third party engaged by Hotel Manager at the Hotel for which a fee is charged to Owner beyond the Base Management Fee.

System Standards   *see* Brand Standards

Tax Withholding   a government requirement for the payer of an item of income to withhold or deduct tax from the payment, and pay that tax to the government. In most jurisdictions, withholding tax applies to employment income. Many jurisdictions also require withholding tax on payments of interest or dividends. In most jurisdictions, there are additional withholding tax obligations if the recipient of the income is resident in a different jurisdiction, and in those circumstances withholding tax sometimes applies to royalties, rent, or even the sale of real estate.

Technical Services Agreement   agreement pursuant to which technical services are provided by Operator to Owner in connection with the Opening of the Hotel.

Technical Services Fee   a fee for the provision of technical and design services, which includes the establishment of the operations, review and approval of designs and plans, and other similar expertise, provided by Manager prior to the Opening of the Hotel or during redevelopment of the Hotel. The Technical Services Fee is typically expressed as a range per Hotel room and is paid during the development or redevelopment period. This is highly variable based on the number of rooms and the level of finishes required under the plans and specifications.

Technology Fee   a fee for services related to all technology, software, help desk and other related functions. This fee is typically 1–3% of Gross Revenue of the Hotel.

Term   term or length of the Operative Agreement, which includes the Initial Term, together with any Renewal Term.

Termination   expiration or earlier termination of the Operative Agreement for any reason.

Termination Fee   any negotiated fee that Owner agrees to pay to Operator on Termination in certain circumstances as set forth in the Operative Agreement. A Termination Fee might be a fixed amount that declines over the Term of the Operative Agreement or a set calculation that reflects the payment to the Operator of a certain multiplier of past management fees paid under the Operative Agreement. For example, the Termination Fee might be $"X" from the Effective Date through the end of the fifth Fiscal Year, $"Y" from the commencement of the sixth fiscal year through the end of the 10th Fiscal Year, etc. Or the Termination Fee might be calculated as the actual Management Fees paid over the previous —one to two Fiscal Years multiplied by a predetermined multiplier. The Termination Fee is typically applicable in all circumstances except where the Operator is in default under the Operative Agreement. The Termination Fee, while considered a Liquidated Damages amount, is often not exclusive of any other remedies that may be available at law or in equity to the Operator.

Termination Upon Sale   right of Owner to terminate Manager upon the sale of the Hotel, and for which Owner must pay Manager a Sale Termination Fee.

Third-Party Financing   financing provided by a source other than the Franchisor. Many Franchisors have relationships with banks or are registered with the Small Business Association in order to expedite the loan process for their Franchisees.

Third-Party Manager   a Hotel Manager that is not affiliated with a Brand.

Trade Name   any name, whether informal (such as a fictitious name or d/b/a) or formal (such as the full legal name of a corporation or partnership) that is used to identify an entity.

Trademark   a word, phrase, symbol, and/or design that identifies and distinguishes the source of the goods of one Party from those of others.

Transfer   any transaction whereby the interest of a Party in the Operative Agreement may be sold, granted, conveyed, leased, assigned, exchanged, transferred, disposed of, encumbered, pledged, charged, mortgaged, hypothecated, given, devised, or bequeathed.

Transferee   Person receiving the interest being Transferred.

Transferor   Person transferring the interest to the Transferee.

TSA   *see* Technical Services Agreement

Uncontrollable Expenses   real estate taxes, utilities, insurance premiums, license and permit fees, fuel costs, and charges provided for in contracts and leases entered into pursuant to the Hotel Management Agreement and in compliance with Applicable Laws, and other expenses that are not within the ability of Manager to control.

Uniform System   *see* Uniform System of Accounts

Uniform System of Accounts   latest revised edition of the Uniform System of Accounts for the Lodging Industry, as published by the Hotel Association of New York City, Inc.

Unit   hotel condominium units, or, in some cases, residential condominium units.

Unit Maintenance Agreement   agreement pursuant to which Manager provides maintenance services to the Unit Owners in connection with the Units.

Unit Owners   owners of Units.

Vouchers   a certificate or other promotional material that may be used as full or partial payment for certain services at the Hotel.

WARN Act   *see* Worker Adjustment and Retraining Notification Act

Worker Adjustment and Retraining Notification Act   Worker Adjustment and Retraining Notification Act, 29 U.S.C. 2101 *et seq*.

Working Capital   funds, as reasonably determined by Manager, for normal day-to-day Hotel operations in accordance with the approved Annual Budget and the terms of the Management Agreement.

Index